FLORA OF TROPICAL EAST AFRICA

ERICACEAE

HENK BEENTJE[1]

Shrubs or trees, usually evergreen. Leaves alternate, opposite or whorled, simple, often very narrow, exstipulate. Flowers actinomorphic or slightly zygomorphic, bisexual, in axillary or terminal racemes or in terminal umbels, bracteate. Calyx and corolla (3–)4–5(–7)-merous; sepals free or fused; corolla with tube and 3–5 lobes; lobes contorted or imbricate. Stamens in two whorls and twice as many as the petals or less often fewer in number, inserted on a disc; filaments free or occasionally connate to corolla at base; anthers opening by terminal or lateral pores, rarely by slits, often with appendages. Carpels 3–5, sometimes one fewer than the number of petals, or only one (the others aborted), forming a (1–)3–5-locular ovary, usually superior; ovules many or less often 1–2 per locule; placentation axile or basal; style hollow. Fruit a capsule, usually loculicidal but sometimes septicidal, or a drupe or berry, rarely a nut.

The Heath family has some 100 genera and 3400 species; cosmopolitan.

Ficalhoa, treated in T.T.C.L.: 193 (1949) as a member of the Ericaceae, is included in Theaceae in FTEA.

1. Leaves > 1 × 0.5 cm; calyx and corolla 5-merous; stamens 9–10
      (subfam. *Vaccinioideae* Endl.) 2
   Leaves < 1.5 × 0.4 cm; calyx and corolla 3–4-merous; stamens
      4–8(–9) (subfam. *Ericoideae*) . . . . **3. Erica**
2. Ovary superior; fruit a capsule . . . . **1. Agarista**
   Ovary inferior; fruit a succulent berry . . . . **2. Vaccinium**

Rhododendron mucronatum (Bl.) G.Don from Japan has been cultivated at Amani, cited in T.T.C.L.: *Greenway* 1695 (not seen)

1. **AGARISTA**

G.Don, Gen. Hist. 3: 837 (1834); Judd in J. Arnold Arbor. 65: 255–342 (1984)

Leucothöe D.Don subgen. *Agarista* (G.Don) Drude in Engler & Prantl, Nat.
Pflanzenfam. IV. 1: 42 (1889)
Leucothöe D.Don sect. *Agauria* A.P.DC., Prodr. 7: 602 (1839)
Agauria (A.P.DC.) J.D.Hooker in G.P. 2: 586 (1876); Sleumer in E.J. 69: 374–394 (1938)

Evergreen shrubs or trees. Leaves alternate to subopposite. Inflorescences in axillary or terminal racemes or panicles; bracts solitary, small (to large and foliaceous), caducous; pedicels with 2(–several) caducous bracteoles. Flowers 5-

[1] Dedicated, with respect, to Olov Hedberg, who has done much to clarify the confused taxonomy in this family (and many other families) in his pioneering works on the Afro-alpine flora.
   I am grateful to Ted Oliver for carefully reading my treatment and for suggesting various improvements; and to Charles Nelson for helping to track down well-hidden taxa names.

merous; calyx of imbricate lobes, articulate with pedicel, persistent in fruit. Corolla cylindrical to urceolate. Stamens 10 in 2 whorls, inserted at base of corolla; filaments flattened, somewhat expanded near base, geniculate; anthers dehiscing by introrse-terminal elliptic pores. Ovary 5-locular, with nectariferous disc; stigma truncate to capitate. Fruit a capsule, erect, not separating from valves at dehiscence; seeds very small, dust-like.

31 species in South America, central Africa, Madagascar, Réunion and Mauritius.

Agarista salicifolia (*Lam.*) G.*Don*, Gen. Syst. 3: 837 (1834); Hedberg & Hedberg in Fl. Eth. 4, 1: 48, fig. 133.2 (2003). Type: Réunion [I. Bourbon], *Commerson* s.n. (P-LAM, holo.; BM, iso.)

Evergreen shrub or tree 0.9–20 m, much branched; bole occasionally large, up to 60 cm across; bark grey to brown, corky, fissured; branchlets glabrous to pubescent with simple hairs, sometimes mixed with glandular hairs, young shoots reddish. Leaves entire, rather leathery, dark green above, whitish to blue-grey and matte beneath, lanceolate, elliptic or oblong, rarely broadly elliptic or ovate, 2–10(–16) cm long, 0.5–4(–7.5) cm wide, base cuneate to subcordate, margins entire, apex rounded to mucronate or acuminate, glabrous except for the midrib underneath; petiole 0.5–1 cm, glabrous or pubescent. Inflorescence of axillary and terminal racemes 5–15 cm long with 15–35 flowers; pedicel 2–6 mm long, pubescent, rarely glabrous; bracteoles 1–1.5 mm long, pubescent. Calyx green to reddish, fused in the lower $^1/_3$; lobes triangular, 2.5 mm long, sparsely pubescent with ciliate margins. Corolla pale green or greenish yellow or yellow to creamy white, less often reddish or white with red streaks or yellow tinged with red at base, obconical to slightly urceolate, 7–10 mm long, 4–5 mm in diameter, with 5 triangular lobes 0.7–0.8 mm long, glabrous. Stamens 9–10, orange, the filaments pubescent and geniculate. Ovary globose, ± 2 mm in diameter, sparsely pubescent; style 1, glabrous, 6–8 mm long; stigma small, capitate. Capsule dark green turning reddish, 4–7 mm in diameter, with persistent style. Fig. 1.

UGANDA. Karamoja District: Kadam, Ilipath, Jan. 1957, *Philip* 812!; Kigezi District: 7 km N of Mafuga Forest Station, Dec. 1971, *Katende* 1472!; Mbale District: Bulambuli, Jan. 1936, *Eggeling* 2445!
KENYA. Northern Frontier District: Nyiro, July 1960, *Kerfoot* 1970!; Meru District: Nyambeni Hills, S of Kirima, Oct. 1960, *Polhill & Verdcourt* 302!; Teita District: Taita Hills, Sagala, May 1985, *Taita Hills Expedition* 31!
TANZANIA. Kilimanjaro, between Bismarck and Horombo Huts, no date, *Mwasumbi & Telecki Expedition* 14066!; Ufipa District: Mbizi Forest Reserve, Oct. 1987, *Ruffo & Kisena* 2812!; Iringa District: Mt Image, Mar. 1962, *Polhill & Paulo* 1662a!
DISTR. U 1–3; K 1–7; T 2–8; widespread in tropical Africa from Cameroon to north-east Africa and south to Zambia; Madagascar, Mascarene Is.
HAB. Dry and moist forest, secondary forest, forest margins, evergreen bushland to woodland or clumped or scattered tree grassland, often on rocky slopes, heath zone; fire-resistant; may be locally common or dominant, belt-forming in Virunga; (1050–)1550–3500 m
USES. Leaves and roots poisonous to man and livestock; bark decoction used by Maasai to aid digestion after overeating meat; good firewood
CONSERVATION NOTES. Widespread; least concern (LC)

SYN. *Andromeda salicifolia* Lam., Encycl. Méth. 1: 159 (1783), as *salicisfolia*; Hook. in Curtis Bot. Mag. 60: t. 3286 (1833)
    A. pyrifolia Pers., Syn. Pl. 1: 481 (1805). Type: Réunion, *Aubert* s.n. (not at BM or L)
    Leucothoe salicifolia (Lam.) DC., Prodr. 7: 602 (1839)
    L. salicifolia (Lam.) DC. var. *pyrifolia* (Pers.) DC., Prodr. 7: 603 (1839)
    Agauria salicifolia (Lam.) Oliv., F.T.A. 3: 483 (1877); Z.A.E.: 509 (1914); Sleumer in E.J. 69: 381 (1938); T.T.C.L.: 189 (1949); K.T.S.: 178 (1961); F.P.U.: 114 (1962); Ross in F.W.T.A. ed. 2, 2: 2 (1963); Verdcourt & Trump, Common Poison. Pl. E.A.: 117 (1969); Letouzey in Fl. Cam. 11: 188 (1970); Hamilton, Uganda For. Trees: 159 (1981); Ross in F.Z. 7(1): 158, t. 26 (1983); K.T.S.L.: 443, fig., map (1994)

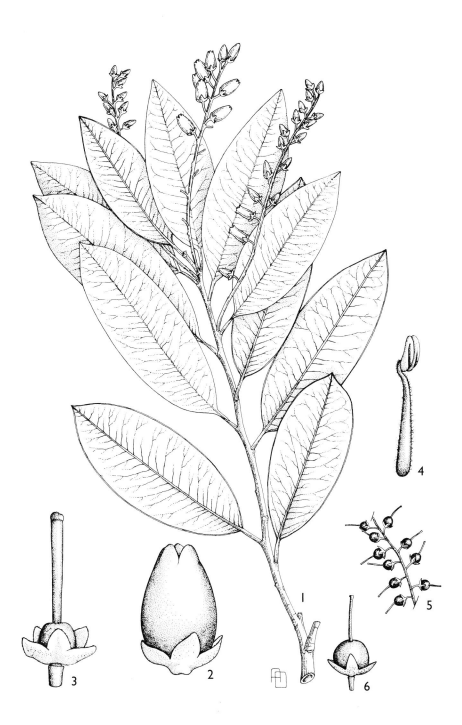

FIG. 1. *AGARISTA SALICIFOLIA* — **1**, habit, × ²/₃; **2**, flower, × 4; **3**, calyx and gynoecium, × 4; **4**, stamen, × 8; **5**, fruiting branch, × ²/₃; **6**, fruit, × 2. All from *Robinson* 5706, and reproduced with permission from Flora Zambesiaca. Drawn by AD.

A. salicifolia (Lam.) Oliv. var. *pyrifolia* (Pers.) Oliv., F.T.A. 3: 483 (1877); Z.A.E.: 509 (1914); T.T.C.L.: 190 (1949)

A. salicifolia (Lam.) Oliv. var. *latissima* Engl. in Hochgebirgsflora Trop. Afr.: 324 (1892); P.O.A. C: 301 (1895); T.T.C.L.: 190 (1949). Type: Tanzania, Kilimanjaro, between Marangu and Machame [Madschame], *Meyer* 312 (B†, holo.)

A. goetzei Engl. in E.J. 30: 369 (1901). Type: Tanzania, Rungwe District: Ngosi/Poroto Crater, *Goetze* 1299 (B†, holo.; BR!, iso.)

A. salicifolia (Lam.) Oliv. var. *pyrifolia* (Pers.) Oliv. forma *adenantha* Sleumer in E.J. 69: 390 (1938); T.T.C.L.: 190 (1949). Type: Tanzania, Njombe District: Likanga on upper Ruhudje, *Schlieben* 223, 1161a (B†, holo.)

A. salicifolia (Lam.) Oliv. var. *intercedens* Sleumer in E.J. 69: 390 (1938); T.T.C.L.: 190 (1949). Type: Tanzania, Pare District: S Pare Mts, Tona, *Peter* 8833 (B†, holo.)

A. salicifolia (Lam.) Oliv. var. *intercedens* Sleumer forma *glandulosa* Sleumer in E.J. 69: 391 (1938); T.T.C.L.: 190 (1949). Type as for *A. goetzei* Engl.

A. salicifolia (Lam.) Oliv. var. *latissima* Engl. subvar. *reducta* Sleumer in E.J. 69: 391 (1938); T.T.C.L.: 190 (1949). Type: Tanzania, Kilimanjaro, above Bismarck Hill, *Peter* 916 (B†, holo.)

A. salicifolia (Lam.) Oliv. var. *pyrifolia* (Pers.) Oliv. subvar. *parvifolia* Brenan, T.T.C.L.: 190 (1949), *nomen nudum*

NOTE. Sleumer in E.J. 69 (1938) recognises a large number of varieties, subvarieties and forms based on leaf size, leaf shape, and the presence/absence of glandular hairs. Many East African specimens have been named var. *latissima*, var. *intercedens*, or var. *pyrifolia* (which should really have been var. *salicifolia*) with various subvars. and formae. This is a common taxon with wide distribution and resistance to fires; with the vast number of specimens now at hand there seem to be no discontinuity in variation. Specimens from large forest trees and those from shrubs in rocky sites can be virtually indistinguishable on a herbarium sheet. The only forms that are almost discontinuous (but not enough to warrant formal status) are:

1. At higher altitudes (above 2900 m) the species is more often represented by a shrub with smaller leaves, and often with reddish flowers.

2. There is a form with long hairs on the branchlets (forma *adenantha* Sleumer) which seems confined to rocky sites by waterfalls and stream-sides, but plants from the same habitat may be glabrous as well.

I follow Ross in F.Z. in regarding the continental African material as a single variable species, without infraspecific splitting.

Eggeling 6513 from **T** 7, foothills of Mt Rungwe, Kiwira River, is a small (30–45 cm high) willow-like shrublet on vertical banks of deep-cut streams – although it is not clear from the label how close to the water – with drooping or creeping branches. Eggeling was certain that it was more than a variety, but leaf and inflorescence characters merge with those of the main population.

Agauria goetzei Engl. in E.J. 30: 369 (1901). Type: Tanzania, upper Konde, Ngozi/Poroto Mts, *Goetze* 1299 (B†, holo.) is presumably this species, but the type has been destroyed. Engler said it differed in the long glandular hairs on the leaf, which would indeed be unusual for this taxon.

2. VACCINIUM

L., Sp. Pl. 1: 349 (1753) & Gen. Pl. ed. 5: 166 (1754)

Small trees or shrubs, evergreen or deciduous. Leaves alternate. Flowers in axillary or terminal racemes, often with bracts, or solitary and axillary, 4–5-merous (5-merous in our species), epigynous. Calyx lobes small, persistent in fruit. Corolla gamopetalous, rotate to urceolate-globose, lobes minute to almost free. Stamens 8–10, anthers at apex with tubular horn opening by terminal pore. Ovary inferior, 4–5-locular; ovules few; style single. Fruit a succulent berry.

200–250 species mainly in Malesia and N temperate zone but extending to mountains in the tropics; 2 African and 1 Madagascar species.

FIG. 2. *VACCINIUM STANLEYI* — **1**, habit, × ²/₃; **2**, leaf, lower surface, × 5; **3**, group of flowers, × 3; **4**, section of flower, × 6; **5**, stamen, × 20; **6**, fruit, × 5. 1, 5 from *Lewalle* 3825; 2 from *Reekmans* 10363; 3–4, 6 from *Purseglove* 2515. Drawn by Juliet Williamson.

Vaccinium stanleyi *Schweinf.* in Sitzungsber. Ges. naturfr. Freunde Berlin 1892: 173 (1892); P.O.A. C: 302, t. 36/d–m (1895); Z.A.E.: 509 (1914). Type: Congo-Kinshasa, Ruwenzori, W side between 2700 and 3400 m, *Stuhlmann* ?s.n. (B†, syn., "several plants, in flower and fruit"; see Note)

Dense low shrub 30–240 cm high (to 4 m at low altitudes in Rwanda); young branches minutely puberulous with short simple hairs but soon glabrescent. Leaves shiny green above, elliptic or slightly ovate, 1–3.3(–5.5) cm long, 0.5–1.7(–2.3) cm wide, base cuneate, margins minutely serrulate with ?glandular teeth, apex acute to very shortly acuminate, glabrous except for the puberulous petiole and the leaf-base which has few puberulous hairs. Flowers 5–15 together in upper-axillary and terminal racemes to 3.5 cm long at anthesis; axes usually glabrous, but occasionally puberulous in proximal parts; bracts ovate, 1–2.2 mm long, to 1.4 mm wide, (to 5.5 × 2.2 mm at fruiting stage), apex acute, margins cilate; pedicel 1.5–3.5(–9) mm long, glabrous, at fruiting stage to 6 mm long and becoming thicker near apex. Calyx lobes green, sometimes tinged with red, ovate-triangular and ± 1 mm long, acute, glabrous. Corolla white, pink or red, campanulate, 2.5–4.2 mm long, 2.5–5.5 mm in diameter at the mouth, connate for more than $^3/_4$, lobes five, 1–1.5 mm long, obtuse, with recurved margins, glabrous. Stamens 10, filaments 0.6–2 mm long, glabrous, anthers 1.1–1.2 mm long, papillose, slightly urceolate at apex. Ovary (?disk) 2 mm across, globose; style 3 mm long, rather thick, slightly widened near apex. Fruit globose, red turning blue-black or purple, 6–7 mm, the remains of the calyx lobes at apex. Fig. 2

UGANDA. Kigezi District: Butahu [Butagu], July 1893/1894, *Scott Elliot* 8007! & Mt Sabinio, Dec. 1930, *B.D.Burtt* 2963! & Mt Mgahinga, Oct. 1947, *Purseglove* 2515!
DISTR. U 2; Congo-Kinshasa, endemic to Ruwenzori
HAB. Montane grassland, giant heath zone on exposed lava ridges; said to form a belt just above the bamboo zone on Mt Sabinio; (2250–)2900–3300 m
USES. Berries edible (fide *Osmaston*)
CONSERVATION NOTES. Data deficient (DD); there is quite a lot of material on the Congolese side, so probably LC.

NOTE. "Wiry herb ... growing below 10' tree heather, just as blueberry creeps below heather on Scottish moors" (*Eggeling* 1111).
P.O.A. C says that the type collection is numbered 2372.
There are resemblances between our species and *Vaccinium exul* Bolus from Malawi and South Africa, but differences in corolla size (2.5–4 mm long with a diameter of 4–5.5 mm in ours, 5–6.5 mm long with a diameter of 3–4.5 mm in *V. exul*) and anthers (1.1–1.2 mm and slightly urceolate in ours, 3–3.5 mm with long horns in *V. exul*) lead me to keep them separate.

3. **ERICA**

L., Sp. Pl. 1: 352 (1753) & Gen. Pl. ed. 5: 167 (1754); E.G.H.Oliv. in S.Afr. J. Bot. 53: 455–458 (1987) & in K.B. 48: 771–780 (1993) & in Contrib. Bolus Herb. 19: 1–483 (2000)

Blaeria L., Sp. Pl. 112 (1753); Hedberg in Nordic J. Bot. 5: 463–467 (1985)
Philippia Klotzsch in Linnaea 9: 354 (1834)
Ericinella Klotzsch in Linnaea 12: 222 (1838)

Trees or shrubs, the shrubs sometimes very small, usually much branched. Leaves in whorls of 3–6 or occasionally spirally inserted, very small, slightly fleshy, the underside of the leaf often with a distinct sulcus. Inflorescence of axillary flowers, in 1–several whorls arranged in elongate racemose to umbellate groups, either at the ends of leafy lateral branches or on very reduced lateral branches or on the main branches. Pedicels short, normally bearing 1 bract (at base, resembling a leaf; on the pedicel; or adjacent to the calyx and resembling a sepal) and 2 opposing bracteoles (in various positions: from basal to appressed to the calyx; rarely missing). Calyx segments 3–4, free or fused, but the situation becomes complicated when bract and/or bracteoles are appressed to the calyx and true sepals in that case may be few or even (as in *E. rossii*) completely absent and replaced by bracts/bracteoles. Corolla hypogynous, globular to tubular, 4-lobed (in ours), the lobes usually shorter than the tube. Stamens 4–8; filaments free, inserted on the hypogynous disc, often geniculate; anthers often with appendages, dehiscing by lateral pores. Ovary 3–4-locular (in ours), placentation axile; style 1, ± expanded at apex. Fruit a loculicidal capsule, contained within the persistent calyx and corolla. Seeds several.

± 600 species, mostly in the Cape, but extending into tropical Africa (usually above 1000 m) and northwards into Europe.

Oliver has studied the genus and its neighbouring genera in great detail and has come to the conclusion that both *Philippia* and *Blaeria* as well as the smaller *Ericinella* cannot be maintained as distinct taxa. *Philippia* was put into synonymy in S. Afr. J. Bot. 53: 455–458 (1987); the other genera are discussed, for instance, in Kew Bull. 48 (4): 771–780 (1993). The separating characters such as bracteoles and number of calyx lobes (*Philippia*) and number of stamens (*Blaeria*) show a range of variation which does not warrant their separate existence. I concur with Oliver and several new combinations have had to be made.

I have not attempted to monograph the tropical African ericas here – a task that is much needed, but my aim is to produce a workable Flora for our area.

Note: in the following key 'sepals' can stand for actual sepals, or for sepals plus bract and/or bracteoles resembling the sepals!

1. Sepals ovate (at least some) and > 0.5 mm wide; stamens
   (5–)6–8(–9) in number 2
   Sepals linear to narrowly ovate and < 0.5 mm wide; stamens
   4 in number 11
2. Leaves in whorls of 4 3
   Leaves in whorls of 3 or nearly so (sometimes in *E. whyteana*
   spirally arranged) 6
3. Leaves usually with long apical gland-topped hair; sepals
   with long simple hairs; near sea level . . . .   1. *E. mafiensis*
   Leaves without apical hair; sepals glabrous or nearly so
   (except *E. benguelensis*, which has short hairs); found
   above 1500 m 4
4. Sepals ± unequal; pedicel and ovary puberulous; stigma
   peltate, > 0.5 mm wide . . . .   2. *E. benguelensis*
   Sepals ± equal; pedicel and ovary glabrous; stigma < 0.5 mm
   wide 5
5. Sepals united for ± 30%; corolla campanulate, 1.5–4 mm
   long . . . .   3. *E. arborea*
   Sepals united for ± 10%; corolla urceolate, 3.5–6 mm long   4. *E. whyteana*
6. Small subshrub, usually less than 50 cm high; corolla
   urceolate, 3.5–6 mm long . . . .   4. *E. whyteana*
   Shrubs or trees; corolla campanulate (rarely urceolate in
   E. trimera or *E. kingaensis*) 7
7. Corolla 3-partite; stamens 6–7; ovary glabrous; stigma >
   0.5 mm in diameter . . . .   6. *E. rossii*
   Corolla 4-partite (and only *E. mannii* with 6–7 stamens, but
   with hairy ovary) 8
8. Stigma flat, circular and peltate, > 0.5 mm in diameter; sepals
   unequal 9
   Stigma capitate, < 0.5 mm across; sepals ± equal (except in
   E. kingaensis) 10

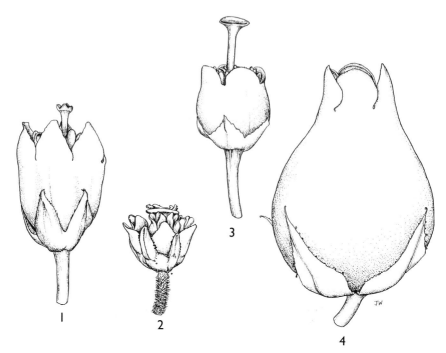

FIG. 3. *ERICA* flowers, all × 12: **1**, *ERICA ARBOREA*; **2**, *ERICA BENGUELENSIS*; **3**, *ERICA TRIMERA* subsp. *ELGONENSIS*; **4**, *ERICA WHYTEANA*. 1 from *Grimshaw* 93/602; 2 from *Drummond & Hemsley* 2960; 3 from *Hedberg* 4529; 4 from *Harvey, Mungai & Vollesen* 6. Drawn by Juliet Williamson.

9. Ovary glabrous; stems with simple hairs; pedicel glabrous; one calyx lobe/bracteole free, the other 3 connate at base .... 5. *E. trimera*
    Ovary puberulous; stems with mixed simple and dendritic hairs; pedicel puberulous (in subsp. *usambarensis*); four calyx lobes connate at base .... 7. *E. mannii*
10. Sepals united for ± 30%; ovary glabrous .... 3. *E. arborea*
    Sepals united for < 10%; ovary hairy .... 8. *E. kingaensis*
11. Flowers and most leaves on short lateral branches with shorter internodes than main stem .... 9. *E. silvatica*
    Most flowers and leaves on numerous erect stems .... 10. *E. filago*

1. **Erica mafiensis** (*Engl.*) *Dorr* in Novon 4: 220 (1994). Type: Tanzania, Rufiji District: Mafia, *Busse* 417 (G, lecto., chosen by Dorr; B†; EA, iso.)

Shrub or small tree 1–5 m high, much branched, the branches erect, grey-black, glabrous, youngest branchlets grey-puberulous. Leaves in whorls of 4, grey, linear-lanceolate, slightly convex above, convex and sulcate beneath, 0.9–4.5 mm long, 0.5–1 mm wide, glabrous or margins with shortly stalked glands, apex subacute and often with gland-topped hair, otherwise glabrous or puberulous; petiole 0.3–0.6 mm, puberulous. Flowers 7–15, in small clumps at branch apices; pedicel 0.5–1 mm long, with sparse but long spreading white simple hairs; bracteoles none. Calyx (apparent calyx; some lobes are really recaulescent bracteoles) 4-merous, three lobes shortly triangular and 1–1.7 mm long, the 4th 1.8–2.2 mm long, leaf-like, green, with apical gland-topped hair, all lobes sparsely hairy with hairs like those on pedicel. Corolla red to brownish, campanulate, 1.5–1.7 mm long, lobes about $\frac{1}{2}$ or $\frac{1}{4}$ of that length, obtuse to rounded, glabrous. Stamens 5–8, usually 7, coherent, filaments flattened, anthers obovoid, 1–1.3 mm long, fissured to halfway. Ovary subglobose, ± 1 mm, puberulous, style 1–2 mm long, stigma peltate, broadly orbicular, 0.7–1 mm in diameter, with 4 raised lines forming a cross on upper surface.

TANZANIA. Rufiji District: Mafia Island, Aug. 1873, *Kirk* s.n.! & Kilindoni, Sep. 1937, *Greenway* 5257!; Pemba: Kinazini, May 1981, *Ruffo* 1679!
DISTR. **T** 6; **P**; not known elsewhere
HAB. Secondary (?fire-induced) vegetation on white sands, where it may be very common to dominant, often forming pure stands; 0–50 m
USES. Friction fire-sticks, brooms, fences
CONSERVATION NOTES. Least concern (LC); locally common on Pemba, even though periodically burned (pers. obs.)

SYN. *Philippia mafiensis* Engl. in E.J. 43: 367, t. 1/a–e (1909); U.O.P.Z: 409 (1949); T.T.C.L.: 194 (1949)

2. **Erica benguelensis** (*Engl.*) *E.G.H.Oliv.* in K.B. 47, 4: 666 (1992). Type: Angola, Huilla, Morro de Lopollo, *Welwitsch* 2560 (B†, holo.; BM!, G, K!, iso.)

Shrub or tree 0.3–8 m high, evergreen; bole gnarled, bark dark brown and peeling off in strips; much branched from near base with ascending branches; branches with poorly developed infrafoliar ridges, densely puberulous to short-pubescent with short simple hairs and usually also with longer glandular hairs to 0.5 mm. Leaves in whorls of 4, appressed to spreading, pale green, straight or slightly incurved, fleshy, flat above, convex and sulcate beneath, narrowly elliptic, 0.8–4.5 mm long, 0.4–0.9 mm wide, margins minutely 'denticulate' with a few glands, apex acute or obtuse, sparsely (glandular-)puberulous on both surfaces or glabrous; petiole 0.3–0.7 mm long, puberulous. Inflorescence with nodding flowers in clusters of 4–12 at the tips of branchlets; pedicel reddish, 1–2.2 mm long, puberulous with minute simple or glandular hairs. Calyx (apparent calyx; some lobes are really recaulescent bracteoles) 4(–5)-partite, green, bowl-shaped, 3 lobes subequal, broadly triangular, 0.8–1.2 mm long, apex acute and slightly thickened, the 4th lobe slightly longer, apex green and slightly leaf-like, all lobes glabrous or puberulous with simple or glandular hairs. Corolla 4(–5)-partite, dark pink to red, fading to brown, rarely whitish, bowl-shaped, 1.1–1.8 mm long, 1.5–2.5 mm in diameter, pubescent outside, the lobes ovate, 0.4–0.8 mm long, apex broadly rounded. At anthesis the red fused stamens surrounding the red capitate stigma may form a connate body to 2 mm in diameter; stamens 8, rarely 5–9 in some flowers, slighly exserted, anthers remaining fused after dehiscence, filaments 0.5–0.7 mm long, anthers 0.6–0.9 mm long. Ovary 4-locular, puberulous, to 0.8 mm long, to 1.5 mm in diameter in fruit; style 0.1–0.3 mm long, stigma peltate, 0.6–1.5 mm in diameter, with centrally 4–5 crossing raised lines above. Fruit splitting in 4 at maturity showing the central keel on the 4 concave valves, with a ridged central axis topped by the capitate stigma. Fig. 3.

Leaves with sparse erect hairs or glabrous . . . . var. *benguelensis*
Leaves with moderately dense appressed hairs on both surfaces . . var. *albescens*

var. **benguelensis**

Leaves with sparse erect hairs, often confined to the upper surface and evanescent; glandular hairs on stems and on apparent margin and apex of leaf, often evanescent. Corolla pale green to green suffused crimson.

UGANDA. Toro District: Ruwenzori, 1 km E of Nyabitaba Hut, Jan. 1969, *Lye* 1381!; Kigezi District. Echuya Forest Reserve, Bufumbira, Apr. 1970, *Katende* 217!; Ankole District: Kalinzu Forest, Aug. 1936, *Eggeling* 3183!

TANZANIA. Pare District: Same, Chabaru Mission, Sep. 1987, *Ruffo* 2524!; Ufipa District: Mbizi Forest, Nov. 1958, *Napper* 1101!; Rungwe District: Kiwara River lower fishing camp, Aug. 1949, *Greenway* 8391!

DISTR. U 2; T 2–7; Congo-Kinshasa, Angola, Zambia, Malawi, Mozambique, Zimbabwe

HAB. Dry rocky hillsides with grassland, bracken or bushland, where it may be locally dominant; also in dry forest edges and clearings, as understorey to woodland; 1500–2500 m

USES. Minor medicinal with the Shambaa (Tanzania)

CONSERVATION NOTES. Least concern (LC); widespread in common habitat

SYN. *Salaxis benguelensis* Engl., Hochgebirgsfl. Trop. Afr.: 328 (1892)
   Philippia benguelensis (Engl.) Britten in Trans. Linn. Soc. Lond. ser. 2, Bot. 4: 24 (1894); Alm & T.C.E.Fr. in K. Svenska Vetensk.-Akad. Handl. ser. 3, 4 (4): 20 (1927); Weimarck in Bot. Not. 1940: 54 (1940), excl. var. *intermedia*; Robyns in F.P.N.A. 2: 21 (1947); T.T.C.L.: 193 (1949); I.T.U. ed. 2: 111 (1951); Brenan in Mem. N.Y. Bot Gard. 8: 493 (1954); K.T.S.: 179 (1961, wrong interpretation); F.P.U.: 114 (1962); F.F.N.R.: 315 (1962); Hamilton, Uganda For. Trees: 80 (1981); R. Ross in Bol. Soc. Brot. ser. 2, 53: 144 (1981) & in F.Z. 7(1): 179 (1983)
   P. milanjiensis Britten [1] in Trans. Linn. Soc. Lond. ser. 2, 4: 24 (1894). Type: Malawi, Mlanje [Milanji], *Whyte* s.n. (BM!, holo.; K, iso.)
   P. holstii Engl., P.O.A. C: 302 (1895). Type: Tanzania, Lushoto District: Kwa Mshusa [Kwa Mshusha], *Holst* 9149 (BM!, K!, iso.)
   P. holstii Engl. var. *glanduligera* Engl., P.O.A. C: 302 (1895). Type: none mentioned, only "Mlalo, Usambara"
   P. stuhlmannii Engl., P.O.A. C: 302 (1895); Z.A.E.: 511 (1914). Type: Congo-Kinshasa, SW Katanda, *Stuhlmann* 2183 & Uganda/Rwanda, Kajonsa [not found], W of Mpororo, *Stuhlmann* 3093 (B†, syn.)
   P. congoensis S.Moore in J. Bot. 57: 213 (1919). Type: Congo-Kinshasa, Katanga, *Kassner* 3352 (BM!, iso.)
   P. kundelungensis S.Moore in J. Bot. 57: 212 (1919). Type: Congo-Kinshasa, Kundelungu, *Kassner* 2769 (BM!, iso.)

var. **albescens** (*R.Ross*) *E.G.H.Oliv.* in K.B. 47, 4: 666 (1992). Type: Tanzania, Mbeya District: Kyimbila, *Stolz* 2059 (BM!, holo.; UPS, iso.)

Leaves with moderately dense white appressed hairs on both surfaces, no glands on stems or leaves. Corolla white.

TANZANIA. Mbeya District: Poroto [Mporoto] Mts, Aug. 1936, *McGregor* 15!; Njombe District: Ukinga, Mwakete, May 1953, *Eggeling* 6582! & Bulongwa, May 1953, *Eggeling* 6621!

DISTR. T 2–7; Zambia, Malawi, Mozambique

HAB. No data for Tanzania, elsewhere by stream amongst rocks; 2100–2250 m

NOTE. These three sheets were the only ones seen for our area

SYN. *Philippia benguelensis* (Engl.) Britten var. *albescens* R. Ross in B.J.B.B. 27: 754 (1957) & F.Z. 7, 1: 180 (1983)

3. **Erica arborea** *L.*, Sp. Pl.: 353 (1753); Oliv., F.T.A. 3: 483 (1877); P.O.A. C: 302 (1895); Z.A.E.: 509 (1914); T.T.C.L.: 192 (1949); I. Hansen in E.J. 75: 37–43 (1950); Pic. Serm. & Heiniger in Webbia 9: 12–28 (1953); A.V.P.: 140 (1957); F.P.U.: 116 (1962); Webb & Rix in Fl. Europ. 3: 7 (1972); Hamilton, Uganda For. Trees: 80 (1981); K.T.S.L.: 444 (1994); U.K.W.F. ed. 2: 170 (1994); Hedberg & Hedberg in Fl. Eth. 4, 1: 46, fig. 133.1 (2003). Type: Hb. Burser. XXV: 24 UPS (Jarvis & McClintock in Taxon 38: 507-514, 1990)

Shrub or tree 0.3–7.5 m high, trunk diameter to at least 5 cm; much-branched, with ascending branches, the branches red-brown; stems pubescent with short smooth hairs and longer dendritic hairs to 1 mm. Leaves in whorls of 3–4, appressed

---

[1] The authority of *P. milanjiensis* has sometimes be cited as Britten & Rendle. In the article both Rendle on his own, Britten on his own, and Britten & Rendle occur; in this case, all that is given is "Welw. mss.", but in the notes the author uses "I" rather than "we", and so I have concluded Britten is the one.

or ascending, needle-like, 2–6.5 mm long, 0.5–1 mm wide, margins minutely denticulate, glabrous; petiole 0.4–0.7 mm long, glabrous. Flowers clustered towards the end of short lateral branches, where their density may give impression of continuous flowers along branches; pedicels 1.5–4 mm long, glabrous or rarely with a few hairs, with bract and 2 ciliolate bracteoles to 1.3 mm long below the middle. Calyx 4-merous, 1.1–1.9 mm long, the lobes ovate, 0.6–1.2 mm long, ciliolate (sometimes with dendritic hairs) to glabrous, saccate at base. Corolla white or pink, campanulate, pendulous, 1.5–4 mm long, widest at the mouth or almost so; stamens 8, included in corolla, anthers bifid and dehiscing by an oblique pore, with two small dorsal/basal appendages; disc present; ovary glabrous, style 0.5–2 mm long, stigma capitate, 0.2–0.5 mm in diameter. Fruit a glabrous capsule. Fig. 3.

UGANDA. Karamoja District: Debasien summit, May 1939, *A.S. Thomas* 2916!; Kigezi District: Muhavura–Mgahuza saddle, Sep. 1946, *Purseglove* 2137! Elgon, Bulambuli, Nov. 1933, *Tothill* 2293!

KENYA. Northern Frontier District: Mt Nyiro, W side, July 1960, *Kerfoot* 2018!; Naivasha District: Kipipiri link road, Jan. 1961, *Lind* 2935!; Mt Kenya, NW slopes, Mar. 1968, *Mwangangi & Fosberg* 568!

TANZANIA. Mbulu/Masai District: Mt Oldeani, Oct. 1988, *Chuwa* 2669!; Mt Meru, Oct. 1948, *Hedberg* 2373!; Kilimanjaro, Shira Plateau near Simba Cave, Sep. 1993, *Grimshaw* 93/602!

DISTR. **U** 1–3; **K** 1–6; **T** 2, 3; Chad (Tibesti), Congo-Kinshasa, Sudan, Ethiopia, Somalia; North Africa and Europe from the Canary Is. and N Spain to the Black Sea region and Saudi Arabia and Yemen

HAB. Dominant in a zone above the forest on many mountains, and probably a pyrophyte; also in open forest, bamboo, grassland, and on upper moorland, often in rocky places; often together with *Stoebe, Artemisia, Faurea* or *Protea* ; (1600–)2000–3900(–4500) m

USES. None recorded; attractive to bees

CONSERVATION NOTES. Least concern (LC)

SYN. *E. acrophya* Fres., Flora 21: 604 (1838). Type: Ethiopia, *Rueppell* s.n. (FR, holo.)
    E. arborea L. subsp. *parviflora* Spirlet in Bull. Seances Acad. Roy. Outre-Mer new ser. 3: 1130 (1957). Type: Congo-Kinshasa, Nyiragongo, *Marlier* s.n. (BR, holo.), **syn. nov.**

NOTE. *Grimshaw* 93/602 notes that some specimens are occasionally stoloniferous.
    The subspecies described by Spirlet was distinguished by characters that fall within the normal variation of this species, and therefore I have made it into a synonym.

4. **Erica whyteana** *Britten* in Trans. Linn. Soc., Bot. 4: 24, t. 5/7–12 (1894); R. Ross in Bol. Soc. Brot. sér. 2, 53: 140 (1981) & in F.Z. 7(1): 167, frontispiece (1983); U.K.W.F. ed. 2: 170, t. 65 (1994). Type: Malawi, Mt Mlanje [Milanji], *Whyte* 59 (BM!, holo.)

Subshrub 10–45(–90)cm high, erect or trailing, sparsely branched; root thick and woody; branchlets red to brown, glabrous, with ridges below the leaves. Leaves spirally arranged or subverticillate in whorls of (3–)4, patent to appressed, green, often flushed with red, subulate, (3.5–)5–12 mm long, (0.6–)0.8–1.5(–2) mm wide, acute and shortly aristate, glabrous, usually with glandular teeth on the margin; petiole ± 1.5 mm long. Flowers solitary in leaf axils forming a short leafy spike on the distal 2–10 cm of branches; pedicel reddish, 2.5–15 mm long, glabrous; bract leaf-like, 2.5–5 mm long, in the basal part of the pedicel, sometimes at base of pedicel; bracteoles around the middle of the pedicel, linear, 0.7–2.5 mm long. Calyx segments 4, almost free, reddish white or pink, broadly ovate, 1.5–3 mm long, acute or acuminate, often reflexed (not in our area), with minute glandular teeth along the margin. Corolla urceolate, pink, often whitish at base, to pinky mauve, sometimes the tips darker, 4–6 mm long, 2.5–4 mm in diameter, the mouth about half that diameter, glabrous, lobes 4, obtuse to rounded, 0.8–1 mm long. Stamens 8, included, anthers with two aristate basal appendages. Ovary globose, 1 mm, glabrous; style 2–3 mm long, glabrous. Fruit reddish, globose, ± 2 mm, hidden within the persistent flower. Fig. 3.

KENYA. Elgeyo District: Cherangani Hills, Lobot, Aug. 1968, *Thulin & Tidigs* 89!; Nyandarua/Aberdare Mts, 0028 S 3650 E, Oct. 1992, *Harvey et al.* 6!; Mt Kenya, above Urumandi, Aug. 1944, *Le Pelley in Bally* 3416!

TANZANIA. Iringa District: Dabaga Highlands, Kibengu, Feb. 1962, *Polhill & Paulo* 1537!;
    Njombe District: N Ukinga, Elton Plateau, May 1953, *Eggeling* 6561! & between Kitulo and
    Matamba, Mar. 1987, *Lovett & Congdon* 1864!
DISTR. **K** 2–4; **T** 7; Malawi, Mozambique, Zimbabwe
HAB. Moorland, usually in swamps or next to streams and often growing in grass tussocks,
    occasionally in bamboo zone or upper forest zone; (1900–)2550–3650 m (?3900 m)
USES. None recorded
CONSERVATION NOTES. Fairly widely distributed; least concern (LC)

SYN. *E. princeana* Engl. in E.J. 43: 363 (1909); T.T.C.L.: 193 (1949). Type: Tanzania, Iringa
        District: Uhehe, Udzungwa [Utschungwe] Mts, *Prince* s.n. (B†, holo.); neotype: Rungwe
        District: Kyimbila, Ukinga, *Stolz* 2605 (K!, neo., BR!, P, UPS, isoneo., chosen by Hedberg)
    E. swynnertonii S.Moore in J.L.S. 40: 128 (1911). Syntypes: Zimbambwe, various localities,
        Swynnerton 648 (BM!, syn.), 1063 (BM!, K, SRGH, syn.), 1064 (BM, K, SRGH, syn.), 1065
        (BM!, syn.)
    E. keniensis S.Moore in Smithson. Misc. Coll. 68 (5): 10 (1917). Type: Kenya, Mt Kenya W
        slope, *Mearns* 1734 (BM!, holo.; S, iso.)
    E. princeana Engl. var. *aberdarica* Alm & Th. Fries in N.B.G.B. 8: 689 (1924). Type: Kenya,
        Nyandarua/Aberdares, Sattima, *Fries & Fries* 2593 (UPS, holo.)
    E. whyteana Britten subsp. *princeana* (Engl.) Hedb., A.V.P.: 142, 297 (1957)

NOTE. There are hardly any differences between the Kenyan and the S Tanzania populations.
    In the F.Z. area sepals are often reflexed, but in our area they are not; *E. princeana* or its
    variety were based on this.
        In Kenya this species is restricted to the Cherangani Hills, the Nyandarua/Aberdare Mts
    and Mt Kenya; in Tanzania it only seems to occur in the Dabaga and Kitulo/Ukinga Mts.

5. **Erica trimera** (*Engl.*) *Beentje* in Utafiti 3, 1: 13 (1990); K.T.S.L.: 444 (1994);
Hedberg & Hedberg in Fl. Eth. 4, 1: 46 (2003). Type: Congo-Kinshasa, Ruwenzori,
3800 m, *Stuhlmann* 2445 (B†, holo.)

Shrub or tree 0.4–12 m high, branched, the branches erect, to 5 mm thick;
branchlets densely leafy, sparsely to densely puberulous with simple and often
glandular hairs, glabrescent. Leaves in whorls of 3 [of 4 according to Engler but I
have not found that condition], porrect or spreading, fleshy, convex beneath, linear,
1–7(–10) mm long, 0.4–1.4 mm wide, margin densely ciliolate with minute hairs or
denticulate with glands (sometimes stalked) or hyaline teeth; or margin glabrous,
apex obtuse, sulcate beneath, glabrous, often shiny with resin exudate; petiole
0.5–1.7 mm long, glabrous or with some marginal hairs. Flowers in clusters of 4–12
at branch ends, 4-merous; pedicels red or pink, 1.5–5.5 mm long, glabrous or with
few small hairs, with the single bract sometimes just below calyx and leaf-like (in
peripheral flowers), sometimes (in central flowers) near the calyx and resembling
the sepals. Calyx (apparent calyx; some lobes are really recaulescent bracteoles) with
4 lobes, 3 (in *elgonensis* 4) connate for 20–30%, ovate or triangular, 0.8–1.4 mm long,
0.7–1 mm wide, obtuse to acute, the 4th free (in *elgonensis* this is the bract) and either
closely against the other lobes or inserted up to 0.7 mm below the other lobes,
narrowly ovate or elongate-triangular with narrow green apex, 1.5–3 mm long,
1–1.1 mm wide; all lobes ciliolate or slightly fimbriate. Corolla pink, red or white,
shortly campanulate with 4 lobes, 1.4–2.6 mm long, 1.6–2.8 mm in diameter, widest
at the mouth, lobes 0.4–1.3 mm long, rounded, glabrous. Stamens 7–8, filaments
1.2 mm long, anthers obovate, 0.7–1.2 mm long, bilobed, connate when young, free
later. Ovary subglobose, ± 1 mm, glabrous; style 1–2.5 mm long; stigma purple,
peltate, circular or slightly lobed, 0.6–1.8 mm in diameter, with 4–5 raised lines near
the centre above. Fruit brownish, subglobose, 1.3–1.4 mm, glabrous.

1. Leaves puberulous on upper and part of lower surface,
      without exudate, usually without glands; **T** 2 . . . .   d. subsp. *jaegeri*
   Leaves at most only with marginal hairs or glands,
      usually shiny from resinous exudate 2

2. Branchlets with stalked glandular hairs only; **U** 2 . . . . .   a. subsp. *trimera*
   Branchlets densely puberulous 3
3. Leaf margin with short hairs and glands; **K** 4 . . . .   b. subsp. *keniensis*
   Leaf margin with glands, but without short hairs (rarely
   short hairs in *kilimanjarica*) 4
4. Corolla 1.5–2.3 mm long, anthers 0.9–1 mm; style > 1 mm
   long; **U** 3, **K** 3 . . . .   c. subsp. *elgonensis*
   Corolla 1.4–2 mm long, anthers 0.7–0.9 mm; style
   0.3–1 mm long; **T** 2 . . . .   e. subsp. *kilimanjarica*

Note: these subspecies are quite close, but the fact that they are absolutely geographically separate has led me to maintain subspecific (rather than varietal) status. Another subspecies, subsp. *abyssinica* (Pic. Serm. & Hein.) ined. occurs in Ethiopia.

a. subsp. **trimera**.

Shrub or tree 3–12 m high; branchlets puberulous with small glandular hairs. Leaves linear, (1–)3–7 mm long, 0.9–1.3 mm wide, margin faintly denticulate with glands (sometimes stalked) or hyaline teeth, glabrous, often shiny with resin exudate; petiole 0.9–1.5 mm long. Flowers with pedicels 2–4.5 mm long, glabrous. Calyx lobes 3, connate for 20–30%, ovate or triangular, 1.2–1.3 mm long, 0.9–1 mm wide, the 4th (actually the bract) free and either closely against the other lobes or inserted up to 0.7 mm below the other lobes, narrowly ovate or elongate-triangular with narrow green apex, 2–3 mm long, 1–1.1 mm wide. Corolla pink or red, 1.6–2.5 mm long, 1.9–2.8 mm in diameter, widest at the mouth, lobes 0.5–1.3 mm long. Stamens 7–8, anthers 0.9–1.2 mm long. Style 0.3–2.5 mm long; stigma peltate, orbicular, 1–1.8 mm in diameter, with 4 raised lines near the centre above; style 0.3–1 mm long.

UGANDA. Ruwenzori, Butahu Valley, Kitandara, Aug. 1933, *Osmaston* 3780! & Mubuku Valley, between Kichuchu and Kabamba, July 1938, *Eggeling* 3793! & 1.5 km SW of Bigo Hut, Jan. 1969, *Lye* 1322!
DISTR. **U** 2; Congo-Kinshasa, endemic to Ruwenzori Mts
HAB. Moorland, open forest; 3000–4000 m
USES. None recorded on specimens
CONSERVATION NOTES. Probably least concern (LC) as the altitudinal range is considerable

SYN. *Philippia trimera* Engl., P.O.A. C: 302, t. 33/a–h (1895); A.V.P.: 144, 298 (1957); R. Ross in B.J.B.B. 27, 4: 745 (1957); Hamilton, Uganda For. Trees: 80 (1981)
    P. longifolia Engl. in E.J. 43: 345 (1909) & in Z.A.E.: 510 (1914); Alm & T.C.E.Fr. in K. Svenska Vetensk.-Akad. Handl. ser. 3, 4 (4): 38 (1927). Type: Congo-Kinshasa, Ruwenzori, *Mildbraed* 2568 (B†, syn.) & 2571 [or 2573, fide Z.A.E. & Alm & Fries] (B†, syn., BR, fragm., not found)
    P. lebrunii Staner in Rev. Zool. Bot. Afr. 22: 243 (1932) & in Ann. Soc. Sci. Bruxelles 53: 159 (1933). Type: Congo-Kinshasa, Ruwenzori, Kiterere Valley, *Lebrun* 4551 (BR, holo.; not found)
    P. humbertii Staner in Ann. Soc. Sci. Bruxelles 53: 160 (1933), *non* Perrier de la Bathie. Type: Congo-Kinshasa, Ruwenzori, W slope, *Humbert* 8915 (BR, holo.; not found, P, iso.)
    P. neohumbertii Staner in Rev. Zool. Bot. Afr. 23: 232 (1933). Type as for *P. humbertii*

b. subsp. **keniensis** (*S.Moore*) *Beentje* in Utafiti 3, 1: 13 (1990); K.T.S.L.: 444 (1994). Types: Kenya, Mt Kenya, anno 1899, *Mackinder* s.n. (BM!, lecto., chosen by Hedberg, 1957)

Shrub 0.4–2 m high; branchlets densely grey-puberulous. Leaves light green, linear to oblong, (1–)2.5–6 mm long, 0.6–1.4 mm wide, margins ciliolate and with glands, viscid; petiole 0.9–1.7 mm long, ciliolate. Flowers: pedicel 2–5.5 mm long, glabrous. Apparent calyx with 3 lobes (2 of which are bracteoles) connate for ± 30%, ovate, 0.8–1.4 mm long, 0.8 mm wide, ciliolate, obtuse to acute, the 4th (the bract) close to other lobes or rarely up to 0.7 mm below them, ovate-lanceolate, 2–2.2 mm long, with greenish apex. Corolla white (2 records), turning brown (2 records), or reddish (1 record), 2.2–2.6 mm long, 2–2.8 mm in diameter; lobes ± 0.5 mm long. Anthers 0.9–1 mm long. Style 1–2 mm long; stigma purple, 1–1.3 mm in diameter, slightly lobed, with a central raised cross or star of lines above.

Kenya. Mt Kenya, above Urumandi, Aug. 1944, *Le Pelley* in *Bally* 3467! & Sirimon Track, Sep. 1963, *Verdcourt et al.* 3770! & Naro Moru Track, Oct. 1967, *Hedberg* 4272!
Distr. **K** 4; endemic to Mt Kenya
Hab. Moist and not too exposed localities, e.g. along streams and below rocks, in the alpine zone, also in the heath zone; (2850–)3000–4500 m
Uses. Not recorded on herbarium specimens
Conservation notes. Probably least concern (LC) as the altitudinal range is considerable

Syn. *Philippia keniensis* S.Moore in J. Bot. 39: 259 (1901); T.T.C.L.: 194 (1949)
  P. hexagona T.C.E.Fries in N.B.G.B. 8: 690 (1924). Type: Kenya, Mt Kenya, W slope, *Fries & Fries* 1359 (UPS, holo.; K, iso.)
  P. trimera Engl. subsp. *keniensis* (S.Moore) Hedb., A.V.P.: 145, 298 (1957); R. Ross in B.J.B.B. 27, 4: 747 (1957); K.T.S.: 180 (1961)

c. subsp. **elgonensis** (*Mildbr.*) *Beentje* in Utafiti 3, 1: 13 (1990); K.T.S.L.: 444 (1994). Type: Uganda, Elgon, 2900 m, June 1920, *Lindblom* s.n. (S, holo.; BR!, fragm., UPS, iso.)

Much-branched shrub or small tree 0.9–3.5 m high; branchlets minutely puberulous. Leaves narrowly ovate, 1.5–3 mm long, 0.4–0.8 mm wide, glabrous or in young leaves with some marginal glands, shiny beneath with resinous exudate; petiole 0.5–0.8 mm long. Flowers with pedicel 3–4 mm long, glabrous; bract 1.6–2.2 mm long, halfway up or fairly close to calyx; calyx 0.8–1.2 mm long, half-connate, 4-lobed with equal lobes, glabrous but sometimes slightly dissected. Corolla whitish, sometimes slightly narrowed towards the mouth, 1.5–2.3 mm, the lobes ± 0.5 mm long. Stamens with anthers 0.9–1 mm long. Style 1.5–2 mm long, stigma peltate, 0.7–1 mm across. Fig. 3.

Uganda. Mt Elgon, Mudange, Aug. 1930, *Saundy & Hancock* 63! & W slope above Butadiri, just W of Crater, Dec. 1967, *Hedberg* 4529! & Sasa trail near Cowthieves' cave, Sep. 1997, *Lye & Pócs* 23079!
Kenya. Elgon, Koitoboss trail, Feb. 1997, *Wesche* 1014!
Distr. **U** 3; **K** 3; endemic to Mt Elgon
Hab. Moorland; 3450–3800 m
Uses. None recorded
Conservation notes. Data deficient (DD)

Syn. *Philippia elgonensis* Mildbr. in Notizbl. Bot. Gard. Berlin 8: 232 (1922)
  Erica elgonensis (Mildbr.) Alm & T.C.E.Fr. in K. Svenska Vetensk.-Akad. Handl. ser. 3, 4 (4): 42 (1927) & in Arkiv Bot. 21A, 7: 15 (1927)
  Philippia keniensis S.Moore subsp. *elgonensis* (Mildbr.) R.Ross in B.J.B.B. 27, 4: 748 (1957); K.T.S.: 180 (1961)
  P. trimera Engl. subsp. *elgonensis* (Mildbr.) Hedb., A.V.P.: 146, 298 (1957)

d. subsp. **jaegeri** (*Engl.*) *Beentje* **comb. nov**. Type: Tanzania, Mbulu/Masai District: Loolmalassin [Lomalasin], *Jaeger* 487 (B†, holo.)

Much branched shrub 1–2 m high or 'large' tree (fide Burtt) in ravines at 3300 m; branchlets puberulous. Leaves linear, 1.5–3.4 mm long, 0.4–0.8 mm wide, margin ciliate with minute hairs or glabrous, puberulous above; petiole 0.5–1 mm long. Flowers with pedicels 3–4 mm long, glabrous or sparsely puberulous. Calyx with 3 ovate or triangular lobes, 0.9–1 mm long, 0.7–0.8 mm wide, the 4th (actually the bract) narrowly ovate or elongate-triangular with narrow green apex, 2–2.2 mm long, 1–1.1 mm wide. Corolla colour not recorded, 1.5–1.8 mm long, 1.8–2 mm in diameter, lobes 0.4–0.7 mm long. Style 1–1.5 mm long; stigma peltate, orbicular, 0.7–0.9 mm in diameter, with 4 raised lines near the centre above.

Tanzania. Arusha District: Mt Meru, Sep. 1932, *Burtt* 4063! & W slopes above Olkakola Estate, Oct. 1948, *Hedberg* 2355!
Distr. **T** 2; endemic to Mt Meru, Mt Loolmalassin
Hab. Dry, open ground, sometimes dominant; 3350–3800 m
Uses. None recorded
Conservation notes. Data deficient (DD)

Syn. *Philippia jaegeri* Engl. in E.J. 43: 368, fig. 2/a–f (1909); T.T.C.L.: 193 (1949)
  P. trimera Engl. subsp. *jaegeri* (Engl.) Hedb., A.V.P.: 145, 298 (1957) pro parte

NOTE. Several specimens have Ross determinavit slips with *Philippia keniensis* subsp. *meruensis* Ross; as far as I know, an unpublished name.

A specimen from Mbulu District: Mt Loolmalassin, Sep. 1932, *Burtt* 4196!, mentioned by Hedberg as subsp. *jaegeri*, has almost glabrous leaves – so is more like Hedbergs own subsp. *kilimanjarica*; it does have puberulous margins, though.

e. subsp. **kilimanjarica** (*Hedb.*) *Beentje* **comb. nov**. Type: Tanzania, Kilimanjaro, Peter's Hut, *Hedberg* 1209 (UPS, holo.; EA, K!, S, iso.)

Shrub 1–3 m, much branched; branchlets grey-puberulous. Leaves narrowly oblong-lanceolate, 1–3.5(–5??) mm long, 0.6–1.3 mm wide, glabrous above, margins glandular and sometimes puberulous with simple short hairs when young, shiny with resinous exudate; petiole 0.6–1.2 mm long. Flowers with pedicel pink, 1.5–3 mm, glabrous; bract variable in position; sometimes just below calyx and leaf-like (in peripheral flowers), sometimes (in central flowers) near the calyx and resembling the sepals, 1.5–1.7 mm long, apex green; sepals pink, subequal, 0.8–1 mm long, 0.7–0.9 mm wide. Corolla pink or white, campanulate, 1.4–2 mm long, 1.6–2.7 mm in diameter, the lobes ± 0.5 mm long. Stamens purple, anthers 0.7–0.9 mm long. Ovary style 0.3–1 mm long; stigma purple, 0.6–1 mm in diameter.

TANZANIA. Kilimanjaro, Peter's Hut to Bismark Hill, Feb. 1934, *Greenway* 3784! & saddle between Kibo and Mawenzi, June 1948, *Hedberg* 1349!; Arusha District: Mt Meru, Little Meru, Jan. 1969, *Richards* 23831!
DISTR. **T** 2; endemic to Kilimanjaro and Mt Meru
HAB. Moorland, heath zone, on dry sloping ground, often burnt and then coppicing (fide *Grimshaw*); (2700–)3150–4550 m
USES. None recorded
CONSERVATION NOTES. Probably least concern (LC) as the altitudinal range is considerable

SYN. *Ericinella mannii* sensu Engl., P.O.A. C: 303 (1895), *non* Hook.f., pro parte
    Philippia jaegeri Alm & T.Fr. in K. Svenska Vetensk. Akad. Handl. 3, 4 (4): 35, pro parte
    P. trimera Engl. subsp. *kilimanjarica* Hedb., A.V.P.: 146, 301 (1957)

NOTE. Hedberg stated this taxon is close to subsp. *trimera*, from which it differs in the shorter flowers and anthers; and to subsp. *jaegeri*, from which it differs in the leaves, hardly pubescent above, resinous beneath.

6. **Erica rossii** *Dorr* in Novon 4: 220 (1994). Type: Kenya, Mt Kenya, *Fries & Fries* 1387 (UPS, holo.; BR, K!, S, iso.)

Shrub or tree 0.3–15 m high, trunk up to 15 cm in diameter with brittle wood; much branched; bark dark brown; branchlets pubescent with minute simple hairs to 0.4 mm long. Leaves in whorls of 3, erect, porrect or spreading, fleshy, narrowly ovate or linear, 1.2–4.8 mm long, 0.3–0.9 mm wide, apex obtuse, sulcate beneath, often shiny with resinous exudate, glabrous but for the margins which have minute hairs and/or glands when young; petiole 0.3–0.7 mm long, glabrous. Flowers 3–9 in fascicles or umbels at branch apices; pedicel pink, 1.2–4 mm long, glabrous or with very few hairs near base. Calyx (apparent calyx; all three lobes are really recaulescent bract + bracteoles) 0.6–1.2 mm long, connate to ¼ or ½, glabrous, three lobes ovate, 0.3–0.7 mm wide, acute, glabrous or minutely ciliolate, one free (the bract), longer and wider at base, 1–2.5 mm long, 0.4–0.6 mm wide, sometimes with green apex. Corolla reddish or pink to greenish white, 3-merous, shortly campanulate, 1.2–1.9 mm long, 1.3–1.9 mm in diameter, widest at mouth, glabrous, lobes 0.4–0.7 mm long, rounded, glabrous. Anthers red, 6–7, 0.9–1 mm long, the tips usually slightly exserted. Ovary 3-locular; style 0.4–2 mm long, glabrous, stigma crimson, flat and peltate, circular, 0.2–0.7 mm long, 0.6–1.3 mm diameter. Fruit loculicidal, 1–1.8 mm long, glabrous.

UGANDA. Ruwenzori Mts, Nyabitaba Hut, Jan. 1967, *Perdue & Kibuwa* 8430!; Kigezi District: Mt Muhavura, Oct. 1948, *Hedberg* 2259! & Echuya Forest Reserve, Apr. 1970, *Katende* 217!
KENYA. Elgon, Apr. 1954, *Tweedie* 1154!; Kiambu District: S Kinangop, Kibata, Nov. 1959, *Kerfoot* 1460!; Mt Kenya, near Naro Moru Lodge, July 1971, *Hedberg* 5022!

TANZANIA. Mbulu District: Mt Hanang, Oct. 1968, *Carmichael* 1525!; Arusha District: Mt Meru, W slopes above Olkakola, Oct. 1948, *Hedberg* 2356!; Kilimanjaro, above Mandera Hut, Sep. 1993, *Grimshaw* 93/691!

DISTR. **U** 2, 3; **K** 3, 4; **T** 2; endemic to Ruwenzori and Virunga Mts, Elgon, the Nyandarua/Aberdares, Mt Kenya, Mt Hanang, Mt Meru and Kilimanjaro; presumably also in Congo-Kinshasa and Rwanda

HAB. Moorland and heath-zone to upper forest and especially clearings and edges, usually with *Ocotea* or *Juniperus* – *Podocarpus*, often locally dominant, and sometimes colonizing clearings in upland forest; (1800–)2300–4050 m

USES. None recorded

CONSERVATION NOTES. Least concern (LC), widespread in a common habitat range

SYN. *Philippia johnstonii* Engl., P.O.A. C: 302 (1895); Z.A.E.: 511 (1914); A.V.P.: 143, 301 (1957); Hamilton, Uganda For. Trees: 80 (1981). Type: Congo-Kinshasa, Ruwenzori, *Stuhlmann* 2374 & 2458 (B†, syn.), *non Erica johnstoniana* Britten; **syn. nov.**
   P. excelsa Alm & T.C.E.Fr. in K. Svenska Vetensk.-Akad. Handl. ser. 3, 4 (4): 41 (1927); T.T.C.L.: 193 (1949); A.V.P.: 143 (1957); K.T.S.: 180 (1961); Hamilton, Uganda For. Trees: 80 (1981)
   Erica excelsa (Alm & T.C.E.Fr.) Beentje in Utafiti 3, 1: 13 (1990); K.T.S.L.: 444, map (1994)

NOTE. See Hemp & Beck on this species as a fire-tolerating component of forests on Kilimanjaro in Phoetocoenologia 31, 4: 449–475 (2001).
   Dale 1000 from **K** 3: Kenya, Equator, Mar. 1957, is close, but has a minute reddish tomentum on stems, leaves and flowers.
   I have added *Philippia johnstonii* to the synonymy. There are no distinguishing characters. Hedberg already considered this taxon very close to *P. excelsa*. Engler in the protologue has *Ph. johnstonii* (Schweinf.) Engl. but there is no indication what the basionym would have been, nor have I been able to trace any.

7. **Erica mannii** (*Hook.f.*) *Beentje* in Utafiti 3, 1: 13 (1990); K.T.S.L.: 444, map (1994). Type: Equatorial Guinea, Bioko, Clarence peak, *Mann* s.n. (K, holo.)– possibly nr 287? (K!)

A shrub or tree 1–10 m tall; much branched, the branchlets densely pubescent with short simple hairs and longer dendritic hairs. Leaves in whorls of 3, ascending to appressed, not curved outwards, linear to narrowly ovate, 2–6 mm long, 0.5–1.3 mm wide, acute, glabrous or the margins with a few simple hairs or glands when young; petioles 0.4–0.7 mm long, glabrous. Flowers in clusters of 3–10 at the tips of the branchlets; pedicels puberulous with short simple hairs or glabrous with occasionally a few dendritic hairs on the lower part, 1.3–5.5 mm long. Calyx (apparent calyx; some lobes are really recaulescent bracteoles) connate for about half-way to $^2/_3$, 0.8–1.5 mm long, the 3 equal lobes ovate, slightly keeled near apex, acute, the 4th (the bract) subequal or up to twice as long as the other 3 and then with green upper part, all lobes connate near base, ciliate, otherwise glabrous. Corolla reddish brown to pale pink (to white, fide Ross in F.Z.), campanulate, 1.2–2 mm long, the 4 lobes 0.5–0.6 mm long, rounded, minutely dissected (?glandular-serrulate?). Stamens usually 6 but 5, 7 or 8 in some flowers of some specimens, anthers free, equalling the corolla or slightly exserted. Ovary 3-locular or 4-locular, pubescent; style exserted, 0.6–2 mm long, the stigma flat, peltate, circular or slightly lobed, 0.6–1.4 mm in diameter, with raised lines on upper surface. Fruit yellow-green tinged reddish.

SYN. *Ericinella mannii* Hook.f. in J.L.S. 6: 16 (1862); Oliv., F.T.A. 3: 484 (1877); P.O.A. C: 303 (1895); Z.A.E.: 511 (1914)
   Philippia mannii (Hook.f.) Alm & T.C.E.Fr. in K. Svenska Vetensk.-Akad. Handl. ser. 3, 4 (4): 37 (1927); R. Ross in F.W.T.A. ed. 2, 2: 2 (1963); Letouzey in Fl. Cam. 11: 201, t. 39/1–6 (1970); R. Ross in Bol. Soc. Brot. ser. 2, 53: 142 (1981) & in F.Z. 7 (1): 178 (1983)

Young branches < 0.4 mm in diameter, with long dendritic
　　hairs; pedicels glabrous or with only 1–2 hairs; ovary 3-
　　locular in most flowers . . . .　　subsp. *pallidiflora*
Young branches > 0.6 mm in diameter, mostly with short
　　dendritic or less often simple hairs; pedicels puberulous
　　with dense hairs; ovary 4-locular in most flowers . . . .　　subsp. *usambarensis*

NOTE. The typical subspecies (subsp. *mannii*) occurs in south-eastern Nigeria, Cameroon
and Bioko.

subsp. **pallidiflora** (*Engler*) *E.G.H.Oliv.* in K.B. 47 (4): 667 (1992). Type: Tanzania, Songea
District: above Bendera, *Busse* 909 (B†, holo.)

Shrub or tree 1–6 m high; DBH to 10 cm, bark rough; youngest branchlets grey, 0.4 mm or
less in diameter, without infrafoliar ridges. Leaves not exceeding 4 mm long, 0.5–1 mm wide;
petiole golden yellow. Pedicels yellowish green, up to 2 mm long, glabrous or with a few
dendritic hairs at the base. Ovary 3–locular in most or all flowers.

TANZANIA. Ufipa District: Ufipa Highlands, 8 km W of Kate, Dec. 1954, *Procter* 318!; Njombe
　　District: Mwakete, Jan. 1957, *Richards* 7827!; Songea District: Mapera village, May 1991, *Ruffo
　　& Kisena* 3239!
DISTR. **T** 4, 7, 8; Congo-Kinshasa, Angola, Zambia, Malawi, Mozambique, Zimbabwe
HAB. *Brachystegia–Uapaca* woodland, giant heath vegetation, wooded grassland, often in rocky
　　situations; may be locally dominant; 1100–3000 m
USES. None recorded
CONSERVATION NOTES. Widely distributed; least concern (LC)
NOTE. May be confused with *E. arborea* when sterile, but with a different distribution.

SYN. *Philippa pallidiflora* Engl. in E.J. 43: 370 (1909); Alm & Fries in K. Svenska Vetensk.–Akad.
　　　Handl., Sér. 3, 4 (4): 40 (1927); T.T.C.L.: 194 (1949)
　　Philippia uhehensis Engl. in E.J. 43: 370 (1909). Type: Tanzania, Iringa District: N Udzungwa
　　　[Utschungwe] Mts, Higulu Plateau, *Goetze* 558 (B†, holo.)
　　Philippia pallidiflora Engl. subsp. *pallidiflora* (Engl.) R.Ross in B.J.B.B. 27: 751 (1957)
　　Philippia mannii (Hook. f.) Alm & T.C.E.Fr. subsp. *pallidiflora* (Engl.) R.Ross in Bol. Soc.
　　　Brot., Sér. 2, 53: 143 (1981)

subsp. **usambarensis** (*Alm & T.C.E.Fr.*) *Beentje* in Utafiti 3, 1: 13 (1990). Type: Tanzania,
Lushoto District: Usambara, *Braun* 2677 (B†, EA, syn.) & 2848 (B†, EA, syn.), same area: *Engler*
1275, *Holst* 38 & 423, *Liebush* s.n. (all B†, syn.), *Eick* 107 (B†, LD fragment, syn.)

Shrub or tree 0.5–8 m high; youngest branchlets 0.5 mm or more in diameter, with
prominent broad infrafoliar ridges. Leaves 2–6 mm long, 0.6–1.3 mm wide. Pedicels up to
5.5 mm long, puberulous with short simple hairs. Ovary 4–locular in most or all flowers.

KENYA. Northern Frontier District: Mt Nyiru, Jan. 1959, *Newbould* 3430!; Machakos District:
　　Nzaui Hill summit, Mar. 1973, *Lawton* 1803!; Teita District: Taita Hills, Vuria summit, Feb.
　　1966, *Gillett et al.* 17095!
TANZANIA. Lushoto District: W Usambara Mts, 2 km NE of Bumbuli Mission, May 1953,
　　Drummond & Hemsley 2458!; Kondoa District: Kinyassi Scarp, Jan. 1928, *Burtt* 956!; Kilosa
　　District: Ukaguru Mts, Mt Mnyera, June 1978, *Thulin & Mhoro* 2806!
DISTR. **K** 1, 4, 6, 7; **T** 2–3, 5–6; Mozambique
HAB. Hilltops, eroded hillslopes, hill-forest clearings and secondary associations; gregarious
　　and locally common to dominant; (750–)1200–2650 m
USES. None recorded
CONSERVATION NOTES. Fairly widely distributed; least concern (LC)

SYN. *Philippia usambarensis* Alm & T.C.E.Fr. in K. Svenska Vetensk.-Akad. Handl. ser. 3, 4 (4): 35
　　　(1927); T.T.C.L.: 194 (1949)
　　Philippia pallidiflora Engl. subsp. *usambarensis* (Alm & T.C.E.Fr.) R.Ross in B.J.B.B. 27, 4: 752
　　　(1957); K.T.S.: 180 (1961)
　　Philippia benguelensis sensu K.T.S.: 179 (1961), *non* (Engl.) Britten
　　Philippia mannii (Hook. f.) Alm & T.C.E.Fr. subsp. *usambarensis* (Alm & T.C.E.Fr.) R.Ross in
　　　Bol. Soc. Brot. ser. 2, 53: 144 (1981) & in F.Z. 7(1): 178 (1983)

8. **Erica kingaensis** *Engl.* in E.J. 30: 370 (1901); Alm & Fries in Arkiv Bot. 21a, 7: 15 (1927); T.T.C.L.: 193 (1949). Types: Tanzania, Njombe District: Ukinga [Kinga] Mts, Mt Djilulu, *Goetze* 920 (B, K!, syn.) & Rungwe District: Rungwe Peak, *Goetze* 1150 (B, BM!, syn.)

Evergreen shrub or tree, 0.5–12 m high, densely branched; branches densely leafy, purplish and glabrous to shortly grey-pilose when young, the hairs ± 0.1 mm long. Leaves in whorls of 3, spreading to porrect, fleshy, flat above, convex beneath, linear or narrowly ellipsoid, 1–5 mm long, 0.5–1.3 mm wide, glabrous or margins ciliolate with dendritic hairs to 0.2 mm long, apex acute or subobtuse, sulcate abaxially; petiole pale, 0.3–1 mm long, with ciliolate margins. Flowers in fascicles or umbels of 2–8 at branch ends, nodding (?always); pedicel mauve to pink, 1–7 mm long, glabrous or puberulous with minute simple or larger dendritic hairs; bract in various positions, sometimes leaf-like and at base of pedicel; bracteoles often opposite near middle of pedicel, narrowly obovate, 1.2–3 mm long, 0.4–0.8 mm wide, concave, denticulate or ciliolate, sometimes reduced or missing. Sepals pink, free almost to base, 4, subequal, submembranaceous, ovate, narrowly ovate to elliptic, 1–3 mm long, 0.8–1.2 mm wide, keeled abaxially, margins with dendritic hairs or minutely ciliolate, apex acute or obtuse. Corolla purple, pink or white with a pink tinge, ovoid or slightly urceolate, 1.7–5 mm long, 1.6–3.6 mm in diameter, the mouth 0.7–2 mm in diameter, shortly 4-lobed, the lobes ± erect, 0.6–1.3 mm long, 0.8–0.9 mm wide, rounded, glabrous. Stamens 7–8, filaments 1–3 mm long, anthers 0.9–1.4 mm long with dorsal basal appendages, the tails 0.2–0.4 mm long, the apex bifid. Ovary 1 mm long, 4-locular, puberulous at apex; style 1.3–4.3 mm long, exserted; stigma capitate (not peltate), 0.3–0.5 mm across. Fruit puberulous, ± globose, opening within the corolla and splitting it.

1. Leaf margins and usually pedicels with dendritic hairs 2
   Leaf margins and pedicels glabrous or with simple hairs;
      U 2, Ruwenzori . . . .   c. subsp. *bequaertii*
2. Sepal margins with dendritic hairs; corolla 2.5–3.6 mm in
      diameter; T 7 . . . .   a. subsp. *kingaensis*
   Sepal margins with simple hairs; corolla 1.7–2.5 mm in
      diameter; U 2, Virunga . . . .   b. subsp. *rugegensis*

a. subsp. **kingaensis**

Evergreen shrub, 0.5–4.5 m high; branchlets densely puberulous with minute simple hairs. Leaves 2–5 mm long, 0.8–1.3 mm wide, margins ciliolate with dendritic hairs to 0.2 mm long; petiole 0.6–1 mm long. Flowers in fascicles or umbels of 2–8 at branch ends; pedicel pink, 1–5 mm long, puberulous with dendritic (rarely simple) hairs. Sepals 2–3 mm long, margins with dendritic hairs. Corolla pink or white with a pink tinge, 3–5 mm long, 2.5–3.6 mm in diameter, the mouth 1.7–2 mm in diameter, the lobes 1–1.3 mm long. Style 1.5–4.3 mm long.

TANZANIA. Mbeya District: Irungu Forest Reserve, along Kitulo road, Nov. 1982, *Magogo* 2303!; Rungwe District: near summit of Mt Rungwe, Oct. 1959, *Procter* 1456!; Njombe District: Ndumbi Forest, Feb. 1954, *Paulo* 249!
DISTR. **T** 7; not known elsewhere
HAB. On slopes and ridges among rocks, where it may form dense stands, scattered in grassland; 1600–3200 m
USES. None recorded
CONSERVATION NOTES. Least concern (LC) as it occurs in a common habitat range, over a reasonable altitude range

b. subsp. **rugegensis** (*Engl.*) *Alm & Fries* in Arkiv Bot. 21 A, 7: 16 (1927). Type: Rwanda, Rukarara, Rugege Forest, *Mildbraed* 979 (B†, holo.; BR!, iso. fragm.)

Shrub 0.5–3 m high; branchlets densely puberulous with minute simple hairs. Leaves 1–4 mm long, 0.7–1 mm wide, the margins with dendritic hairs with a short body to 0.1 mm long; petiole 0.6–0.9 mm long. Flowers 5–7 in umbels at branch apices; pedicel purple, 4–7 mm long, densely puberulous with dendritic or less often simple hairs. Sepals 1.5–2.5 mm long, 0.8–1.2 mm wide, ciliolate. Corolla pale pink, 2.8–3.5 mm long, 1.7–2.5 mm in diameter, mouth 1–1.7 mm across, the lobes 1–1.2 mm long, 0.8–0.9 mm wide. Style 2–3.5 mm long.

UGANDA. Kigezi District: Behungi, Apr. 1927, *Linder* 2579! & Kanaba Pass, Feb. 1945, *Greenway & Eggeling* 7120! & Echuya Forest Bog, Apr. 1970, *Lye & Katende* 5339!
DISTR. **U** 2; Congo-Kinshasa, Rwanda, Burundi
HAB. Grassland and bamboo associations, forming local patches, also swamp edges; 1800–2400 m
USES. None recorded
CONSERVATION NOTES. Least concern (LC) as it occurs in a common habitat range, over a reasonable altitude range

SYN. *E. rugegensis* Engl. in E.J. 43: 345 (1909); Z.A.E.: 510 (1914)
    E. linderi Mildbr. in Journ. Arnold Arboretum 11: 51 (1930). Type: Uganda, Kigezi District: E of Behungi, *Linder* 2579 (B†, holo.: A, K!, iso.) (see also Note)
    E. kingaensis Engl. subsp. *multiflora* Spirlet in Bull. Seances Acad. Roy. Outre-Mer new ser. 3: 1135 (1957). Type: Congo-Kinshasa, Biega, *Marlier* s.n. (BR, holo.), **syn. nov.**
    E. kingaensis sensu Hamilton, Uganda For. Trees: 80 (1981) pro parte

NOTES. At Kew *Linder* 2579 was placed in a type folder, with the note "*Erica kingaensis* Engl. subsp. *linderi* Mildbr. n.var."; I have been unable to trace this name and believe it to be an unpublished one. The specimen is ± intermediate between the two Ugandan subspecies.
    The subspecies described by Spirlet was distinguished by characters that fall well within the range of variation of subsp. *rugegensis*. I have made it into a synonym.

c. subsp. **bequaertii** (*De Wild.*) *R.Ross* in Ann. Mag. Nat. Hist. ser. 12, 9: 95 (1956). Type: Congo-Kinshasa, Ruwenzori, Lanuri Valley, *Becquaert* 4546 (BR!, holo.; iso.)

Shrub or tree 1.5–15 m high; branches glabrous to minutely puberulous. Leaves 1–4 mm long, 0.5–0.8(–1) mm wide, glabrous or with several minute hairs or teeth on the margins; petiole 0.3–0.9 mm. Flowers 2–5 in fascicles at the ends of branches or on short side-branches; pedicel mauve to pink, 1.5–3 mm long, glabrous or puberulous with minute simple hairs. Sepals 1–1.6 mm long, minutely ciliolate. Corolla purple, pale pink or white, 1.7–3.5 mm long, 1.6–2.4 mm in diameter, the mouth 0.7–1.6 mm across, the lobes 0.6–1 mm long. Style 1.3–3 mm long.

UGANDA. Ruwenzori: Mubuku Valley, July 1938, *Eggeling* 3791! & same loc., no date, *Hamilton* 781! & Butahu Valley, Kitandara, Aug. 1953, *Osmaston* 3782–3786!
DISTR. **U** 2; Congo-Kinshasa, endemic to Ruwenzori
HAB. Drier parts of *Sphagnum* bogs, heath vegetation on ridges; 2400–3500 m
USES. None recorded
CONSERVATION NOTES. Data deficient (DD)

SYN. *E. bequaerti* De Wild., Pl. Bequaert. 4: 290 (1927)
    E. butaguensis De Wild., Pl. Bequaert. 4: 291 (1927). Type: Congo-Kinshasa, Ruwenzori, Butagu Valley, *Becquaert* 3632 (BR!, holo.; iso.), **syn. nov.**
    E. ruwenzoriensis Alm & T.C.E.Fr. in Arkiv Bot., Stockh. 21A, 7: 17, fig. 4a–b (1927); Staner in Ann. Soc. Sci. Brux. 53: 155 (1933). Type: Congo-Kinshasa, Ruwenzori, Butagu valley, *Mildbraed* 2554 (B, syn.), 2558 (B, BR!, syn.), 2567 (B, syn.); Mubuku valley, *Kassner* 3118 (B, K, syn.), *Scott Elliot* 7985, 7986, 7999 (K, syn.)
    E. kingaensis Engl. subsp. *leleupiana* Spirlet in Bull. Seances Acad. Roy. Outre-Mer new ser. 3: 1132 (1957). Type: Congo-Kinshasa, Vutumbere, *Celis* s.n. (BR, holo.), **syn. nov.**
    E. kingaensis sensu Hamilton, Uganda For. Trees: 80 (1981) pro parte

NOTE. The subspecies described by Spirlet was distinguished by characters that fall well within the range of variation of subsp. *bequaertii*. I have made it into a synonym.

9. **Erica silvatica** (*Engl.*) *Beentje* **comb. nov.** Type: Tanzania, Kilimanjaro, between Marangu & Machame, *Meyer* 343 (B†, holo.; UPS, iso.)

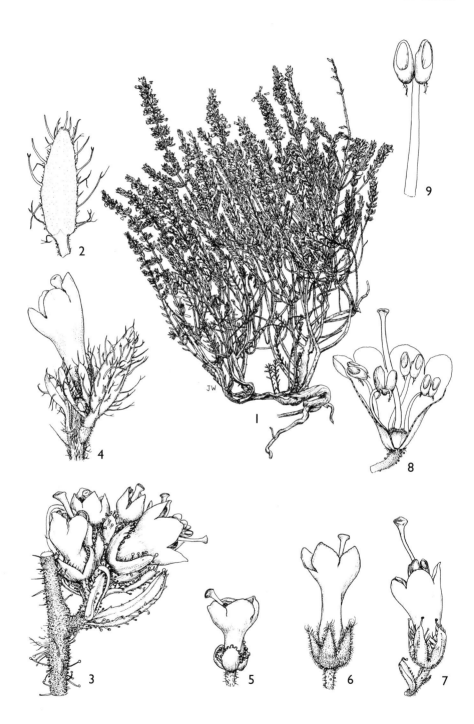

Fig. 4. *ERICA SILVATICA* — **1**, habit, × ²/₃; **2**, leaf upper surface, × 12; **3**, group of flowers, × 12;
4–7, flower side views, × 12; **8**, opened flower, × 12; **9**, stamen, × 24. 1, 2, 4 from *Greenway
& Kanuri* 13257; 3 from *Gilbert Rogers* 570; 5 from *Hedberg* 1164; 6 from *Grote* 3869; 7–9 from
Hedberg 1335. Drawn by Juliet Williamson.

Wiry dwarf shrub 7–40 cm high (rarely to 1 m, and twice reported to 1.8 m, in sheltered situations at lower altitudes), much branched, erect or spreading, with short densely leafy side branches 0.5–5 cm long and long terminal flowering shoots; stem brown to red; branchlets minutely pubescent with simple hairs, intermixed with sparse dendritic hairs 0.3–1.2 mm long, sometimes intermixed with dense glandular hairs 0.1–1.2 mm long which may or may not have minute side-branches near their base. Leaves in whorls of 3(–4), rarely with spirally inserted leaves, porrect or spreading (less often, in moister situations), fleshy, narrowly elliptic or lanceolate, 0.9–6 mm long, 0.3–1.3 mm wide, sulcate beneath, acute, indument variable from sparse to dense, and variable from glandular hairs only to a mixed pubescence of simple and branched hairs, or ciliate with only marginal hairs or stalked glands, or glabrous with only a branched apical hair to 0.6 mm; petiole 0.1–0.4 mm long, usually puberulous. Flowers on small side branches, forming apparent verticillasters 1–15 cm long, also some solitary and axillary to distant leaves; pedicel 0.2–1.7 mm long, puberulous or less often glabrous bract leaf-like, 0.6–1.6 mm long; bracteoles reduced or missing. Sepals 4, green or reddish, ± equal, narrowly ovate or linear, 0.5–2 mm long, 0.2–0.5 mm wide, apex acute, gibbous, ciliate with simple and branched hairs or stalked glands. Corolla pink, less often whitish or violet, 4-merous, infundibuliform or shallowly cup-shaped, 0.8–4.5 mm long, diameter at the mouth 0.6–1.5 mm, the lobes 0.3–0.7 mm long, rounded to subacute. Stamens dark purple, 4(–5), with filaments 0.2–3.5 mm long; anthers partly exserted, 0.2–0.7 mm long, with basal tails 0.1–0.5 mm long or absent. Ovary globose or slightly 4-lobed, puberulous; style red, 0.1–5 mm long, stigma red, capitate, 0.1–0.3 mm in diameter. Fruit globose, ± 1 mm, puberulous, first widening, then splitting the lower part of the corolla tube. Fig. 4.

UGANDA. Karamoja District: Sukdek, summit of Mt Moroto, Apr. 1963, *J. Wilson* 1335!; Mt Elgon, just W of pass into Crater, Dec. 1967, *Hedberg* 4535! & Kimilili trail, 3400 m, Sep. 1997, *Wesche* 1814!

KENYA. Cherangani Hills, Flat Top, Dec. 1959, *Bogdan* 4984!; Nyandarua/Aberdare Mts, Queen's Falls, Dec. 1972, *Kokwaro* 3242!; Nakuru District: Mau forest, Bondui, Jan. 1946, *Bally* 4924!

TANZANIA. Kilimanjaro, Shira Cone, Sep. 1993, *Grimshaw* 93/605!; Morogoro District: Uluguru, Lukwangule Plateau, Mar. 1955, *Semsei* 1979!; Mbeya District: Maniswela Hills, Apr. 1983, *Leliyo* 402!

DISTR. **U** 1–3; **K** 2–5; **T** 2–4, 6–8; Guinea and Ivory Coast (Nimba Mts), Cameroon, Equatorial Guinea (Bioko), Sudan, Ethiopia, Rwanda, Burundi, Angola, Malawi, Zimbabwe

HAB. Short grassland, rock outcrops, stony hillslopes and -crests, moorland, heath zone and bamboo clearings, less often in swamp edges and forest edges; may be locally common or co-dominant (e.g. around Horombo Hut on Kilimanjaro, on the Lukwangule Plateau); (1350–)1800–4000(–4500) m

USES. None recorded on herbarium specimens

CONSERVATION NOTES. Least concern (LC) as it occurs in a common habitat range, over a wide altitude range

SYN. *Blaeria condensata* A.Rich., Tent. Fl. Abyss. 2: 13 (1851), *non Erica condensata* Benth. Type: Ethiopia, Silke Mts, *Schimper* II.667 (BM!, BR!, iso.)
  B. spicata A.Rich., Tent. Fl. Abyss. 2: 13 (1851), *non Erica spicata* Thunb.; F.T.A. 3: 484 (1877); P.O.A. C: 303 (1895); T.T.C.L.: 191 (1949); Pic.Serm. & Heiniger in Webbia 9: 36, fig. 9 (1953); Letouzey, Fl. Cam. 11: 191 (1970); Hedberg in Nordic J. Bot. 5: 463–467 (1985). Type: Ethiopia, Mt Bachit, *Schimper* 749 (BM!, BR!, K!, iso.)
  B. silvatica Engl., Hochgebirgsfl. Trop. Afr.: 326 (1892); P.O.A. C: 303 (1895); T.T.C.L.: 191 (1949). Basionym of *Erica silvatica*
  B. bugonii Engl., Hochgebirgsfl. Trop. Afr.: 327 (1892); T.T.C.L.: 192 (1949). Type: Angola, Huila, *Welwitsch* 2559 (BM!, K!, iso.)
  B. glutinosa K. Schum. & Engl. in Hochgebirgsfl. Trop. Afr.: 327 (1892); P.O.A. C: 303 (1895); T.T.C.L.: 192 (1949). Type: Tanzania, Kilimanjaro, *Meyer* 100 & 235 (B†, syn.)
  B. johnstonii Engl., Hochgebirgsfl. Trop. Afr.: 326 (1892); P.O.A. C: 303 (1895); Alm & T.C.E.Fr. in Acta Hort. Berg. 8 (8): 251 (1924); T.T.C.L.: 190 (1949); A.V.P.: 148, 301, fig. 31–33 (1957); U.K.W.F. ed. 2: 170, t. 65 (1994). Type: Tanzania, Kilimanjaro, *Johnston* 17 (BM!, K!, UPS, iso.) – note: there was no number in the protologue but there is little doubt it is this specimen

B. meyeri-johannis K. Schum & Engl. in Hochgebirgsfl. Trop. Afr.: 326 (1892); P.O.A. C: 303 (1895); T.T.C.L.: 192 (1949). Type: Tanzania, Kilimanjaro, *von Hoehnel* 192 (B†, syn.) and *Meyer* 96, 97 & 206 (B†, syn.; fragm. of 206 at UPS)

B. spicata Hochst. var. *mannii* Engl., Hochgebirgsfl. Trop. Afr.: 325 (1892). Type: Cameroon, Mt Cameroon, *Mann* 1280 (K!, lecto., chosen by Cheek 1997)

B. spicata Hochst. var. *patula* Engl., Hochgebirgsfl. Trop. Afr.: 325 (1892); T.T.C.L.: 191 (1949). Type: Malawi, near Blantyre, Shiré Highlands, *Last* s.n. (K!, holo.)

B. whyteana nomen nudum in Engl. P.O.A. A: 132 (1895)

B. kingaensis Engl. in E.J. 30: 370 (1901); Alm & T.C.E.Fr. in Acta Hort. Berg. 8, 8: 247 (1924); T.T.C.L.: 191 (1949); R. Ross in Bol. Soc. Brot. ser. 2, 53: 123 (1981) & in F.Z. 7(1): 171, t. 29 (1983), *non Erica kingaensis* Engl. (1901). Type: Tanzania, Njombe District, Ukinga Mts, Kipengere, *Goetze* 957 (B†, holo.; BM!, BR!, Z, iso.), **syn. nov.**

B. subverticillata Engl. in E.J. 30: 371 (1901), *e descr.* Type: Tanzania, Njombe District: Livingstone Mts, Masuanu [Masuamu] Mt, *Goetze* 826 (B†, holo.; BR!, iso.)

B. breviflora Engl. in E.J. 43: 364 (1909); Z.A.E.: 511 (1914); Alm & T.C.E.Fr. in Acta Hort. Berg. 8 (8): 259, t. 7–8e, t. 14i (1924); T.T.C.L.: 192 (1949); U.K.W.F. ed. 2: 170 (1994). Type: Tanzania, Lushoto District: W Usambara, Sakare–Manka, *Engler* 1031 (B†, syn.); Kwai, *Eick* 244 (B†, syn.); Lutindi, *Liebusch* s.n. (B†, syn.); Rwanda, Rukarara, Rugege, *Mildbraed* 1004, 1005 (both B†, syn.), **syn. nov.**

B. breviflora Engl. var. *ulugurensis* Engl. in E.J. 43: 364 (1909). Type: Tanzania, Morogoro District: Uluguru, SW, upper margin of 'Rodungsgebietes' at 620 m, *Stuhlmann* 9320 (B†, syn.); Lukwangulu, *Stuhlmann* 9157a (B†, syn.)

B. glanduligera Engl. in E.J. 43: 366 (1909); T.T.C.L.: 190 (1949). Type: Tanzania, Kilimanjaro, above Johannesschlucht, *Volkens* 1170 (B†, holo.; BM, K!, UPS, iso.)

B. keilii Engl. in E.J. 43: 365 (1909), *e descr.* Type: Burundi, Mt Luhona, *Keil* 269 (B†, holo.)

B. kiwuensis Engl. in E.J. 43: 346 (1909); Z.A.E.: 511 (1914); Alm & T.C.E.Fr. "emend." in Acta Hort. Berg. 8, 8: 246, t. 12/a–b, t. 14/c–f (1924); T.T.C.L.: 192 (1949). Type: Rwanda/Congo-Kinshasa, SE of Karisimbi, *Mildbraed* 1569 (B†, holo.; BR!, iso., fragment)

B. mannii (Engl.) Engl. in E.J. 43: 366 (1909); Z.A.E.: 511 (1914); R. Ross in F.W.T.A. ed. 2, 2: 2 (1963), *non Erica mannii* (Hook.f.) Beentje

B. patula (Engl.) Engl. in E.J. 43: 364 (1909); Alm & T.C.E.Fr. in Acta Hort. Berg. 8, 8: 261 (1924); T.T.C.L.: 192 (1949), *non Erica patula* Klotzsch (1838) *nec Erica patula* Benth. (1839). Types: Malawi, near Lungwe, *Stolz* 52; S Nyika, *Whyte* s.n.; Shire Highlands, *Buchanan* 1478 (BM!, K, syn.); Blantyre, *Last* s.n.

B. tenuifolia Engl. in E.J. 43: 365 (1909). Type: Malawi, S Malawi, Maruku Plateau, *Whyte* 276 (K!, holo.)

B. patula (Engl.) Engl. var. *aberdarica* Alm & T.C.E.Fr. in Notizbl. Bot. Gart. Mus. Berlin 8: 694 (1924); T.T.C.L.: 192 (1949). Type: Kenya, Nyandarua/Aberdares, upper bamboo zone, *Fries & Fries* 2391 (UPS, holo.; B†, K!, iso.)

B. breviflora Engl. var. *aberdarica* (Alm & T.C.E.Fr.) Alm & T.C.E.Fr. in Acta Hort. Berg. 8, 8: 260 (1924)

B. granvikii Alm & T.C.E.Fr. in N.B.G.B. 8: 693 (1924). Type: Uganda/Kenya, Elgon, *Granvik* s.n. (S, holo.; UPS, iso.)

B. keniensis Alm & T.C.E.Fr. in N.B.G.B. 8: 694 (1924). Type: Kenya, Mt Kenya, *Fries & Fries* 397 (UPS, holo.; K!, iso.)

B. kilimandjarica Alm & T.C.E.Fr. in Acta Hort. Berg. 8 (8): 259 (1924); T.T.C.L.: 192 (1949), as *kilimandscharica*. Type: Tanzania, Kilimanjaro, *Grote* 3873 (B†, EA, K! mixed gathering, UPS, syn.) & 3914 (B†, EA, K!, syn.)

B. patula Engl. var. *tenuis* Alm & T.C.E.Fr. in Acta Hort. Berg. 8, 8: 261 (1924). Type: Uganda, Mt Elgon, *Dummer* 3503 (B†, holo.; K!, UPS, iso.)

B. stolzii Alm & T.C.E.Fr. in Acta Hort. Berg. 8, 8: 246 (1924); T.T.C.L.: 191 (1949). Type: Tanzania, Rungwe District: Kinga Mts, 'Mwalakola' [I read this as Mwakalila = Mwakaleli, Rungwe], *Stolz* 2111 (B†, BM!, S, UPS, syn.) & 2601 (B†, BM!, K!, syn., MO, !on web), **syn. nov.** is a variant with dense stalked-glandular hairs intermixed on stem and leaves, the leaves also puberulous with minute simple hairs, otherwise similar in all respects.

B. tenuis Alm & T.C.E.Fr. in Notizbl. Bot. Gart. Mus. Berlin 8: 695 (1924)

B. tenuipilosa Alm & T.C.E.Fr. in Acta Horti Berg. 8: 246 (1924); Pic.Serm. & Heiniger in Webbia 9: 40, fig. 10 (1953). Type: Cameroon, Bamboutos Mts, *Ledermann* 1625 (B†, holo.; UPS, lecto., chosen by Pic.Serm. & Heiniger)

B. friesii Weimarck in Bot. Notiser 1940: 60 (1940). Type: Zimbabwe, foothills of Inyangani Mts, *Norlindh & Weimarck* 5027 (BM!, BR!, K, LD, iso.)

B. sphagnicola Sleumer in F.R. 45: 15 (1938); T.T.C.L.: 191 (1949). Type: Tanzania, Pare District: S Pare Mts, Tona Swamp near Wudee, *Peter* 12027 (B†, holo.; K!, iso.), **syn. nov.**

B. sphagnicola Sleumer forma *pubescens* Sleumer in F.R. 45: 16 (1938); T.T.C.L.: 191 (1949). Type: Tanzania, Pare District: S Pare Mts, Tona Swamp near Wudee, *Peter* 41447a (B†, holo.; K!, iso.), **syn. nov.**

B. sphagnicola Sleumer forma *pseudobreviflora* Sleumer in F.R. 45: 16 (1938); T.T.C.L.: 191 (1949). Type: Tanzania, Pare District: S Pare Mts, Tona Swamp, *Peter* 11992 pro parte maiore (B†, holo.; K!, iso.), **syn. nov.**

B. guguensis Pic.Serm. & Heiniger in Webbia 9: 44 (1953). Type: Ethiopia, Gugu, *Milchersich* 77 (FT, holo.), **syn. nov.** ex descr.

B. johnstonii Engl. subsp. *keniensis* (Alm & T.C.E.Fr.) Hedb., A.V.P.: 149, 306 (1957)

B. spicata A.Rich. subsp. *mannii* (Engl.) Wickens in K.B. 27: 513 (1972)

Erica tenuipilosa (Alm & T.C.E.Fr.) Cheek in Kew Bull. 52 (3): 753 (1997), **syn. nov.** Type as for *B. spicata*

Erica tenuipilosa (Alm & T.C.E.Fr.) Cheek subsp. *tenuipilosa*, **syn. nov.**

Erica tenuipilosa (Alm & T.C.E.Fr.) Cheek subsp. *spicata* (A.Rich.) Cheek in Kew Bull. 52 (3): 754 (1997); Hedberg & Hedberg in Fl. Eth. 4, 1: 47 (2003), **syn. nov.**

NOTE. This group was in utter confusion until Hedberg, in A.V.P. (1957), made a critical study of a large amount of material from the afro-alpine zone and showed variation (between what were six species) is continuous. He decided to use the name *Blaeria johnstonii*. I have followed him gratefully, and gone further in uniting the subspecies he proposed; subsp. *keniensis* was distinguished by having rather more spaced leaves, but Hedberg already indicated that intermediates occurred and that some specimens were only referable to subspecies by locality. Since Hedberg's treatment more intermediate collections have been made.

I have also added to the synonymy many taxa that Hedberg did not treat, as they occur at lower altitudes; their measurements and descriptions merge gradually with the those of the upper group. Every attempt to distinguish (and write keys to) varieties has failed, though I have tried hard to distinguish groups based on such characters as the shallowly cup-shaped corolla ('*patula*'). Therefore I treat this as a widespread, very variable taxon.

Variation can be great in the type and degree of indument: this can consist of only stalked glandular hairs (simple or with side branches near base) to a mixture of minute simple hairs and dendritic hairs; the hairs can be sparse or dense, and on the leaves they can be distributed all over or restricted to margins and apex. Sepal shape can vary from linear to narrowly triangular to slightly spatulate; and the shape of the corolla is what most 'species' were based on, varying from less than 1 mm and obconic to shallowly cup-shaped at one end of the range (*patula*) to over 4 mm and distinctly tubular and infundibuliform (*johnstonii* and *spicata*).

At lower altitudes there is a tendency to smaller flowers with concomitant smaller anthers and style, but it is all very gradual and with intermediates for all characters. In this I agree with and follow Ross in Bol. Soc. Brot. ser. 2, 53: 123–149 (1981), who discusses the F.Z. representatives of this group. He studied what he calls *Blaeria kingaensis* (including *subverticillata*, *patula*, *tenuifolia*, *stolzii* and the Zimbabwean *B. friesii*); and reported this is a species which encompasses great variation in indument type as well as corolla length, with ± continuous variation, and no recognizable subgroups. Ross sees the corolla as varying from 1–3 mm, with the shortest ones bowl-shaped without a basal tubular portion. Ross states *Blaeria johnstonii* differs from his *B. kingaensis* in having leaves with a smaller maximum size that are usually incurved; also, though the marginal hairs on their leaves are always < 0.5 mm long, they normally bear side branches on their lower part (these side branches are only rarely found in the FZ area and in SW Tanzania, and then in hairs > 0.5 mm long). I have been unable to uphold this distinction.

Several synonyms are slightly suspect, as the types have been destroyed. From the description in the protologue, *B. subverticillata* belongs here. From the description in the protologue and Engler's notes thereto I also believe *B. breviflora* is a synonym.

The hitherto separate northern taxa, *Blaeria spicata* = *Erica tenuipilosa*, fall into line as well. The only differences with the more southern specimens are a generally slightly longer calyx and corolla, but while extreme specimens can be separated easily, there is again a whole host of intermediates. While the epithet is the oldest available within *Blaeria*, it has been taken up in *Erica* and so cannot be used.

Among the oldest available names is the one used by Hedberg, *Blaeria johnstonii*, so the species really should be called *Erica johnstonii*. However, there already exists an *Erica johnstoniana* Britten; recommendation 23A.2 of the Code says such similarity should be avoided; hence, I have chosen to use a name from among the oldest ones available, in fact as old as Hedbergs chosen epithet. "silvatica" is not a very appropriate name, but the alternatives were names with types destroyed or only known from fragments.

Blaeria patula var. *minima* Brenan; type: Malawi, Nyika Plateau, *Brass* 17291 (K!, holo.) – looks slightly different; it has flat leaves, not succulent and recurved as usual, the flowers are solitary and axillary and spaced; the corolla is shallowly cup-shaped; stamens 5; style 0.3 mm. This would be in the 'patula' series of the species but for the strange leaves.

10. **Erica filago** (*Alm. & T.C.E.Fr.*) *Beentje* **comb. nov.** Type: Kenya, Nyandarua/ Aberdares, Sattima, *Fries & Fries* 2894 (UPS, lecto., B†, K!, iso., chosen by Hedberg)

Dwarf shrub 5–40 cm high (to 50 cm fide F.Z.), erect; stem red, branched from base with short leafy branches; branchlets puberulous with minute simple and/or stalked glandular hairs to 1 mm long, the glandular hairs sometimes (especially on Mt Elgon) with side-branches near base, glabrescent. Leaves greygreen, dense, in whorls of 3 or 4, fleshy, narrowly ovate or narrowly lanceolate, 2–5 mm long, 0.7–1.2 mm wide, sulcate beneath, apex subobtuse, densely grey-puberulous with minute simple and dendritic and/or stalked glandular hairs to 1 mm long, apex with large dendritic hair, ciliate with dendritic hairs; petiole 0.3–0.5 mm long, puberulous. Flowers in axillary fascicles of 2–6, these groups verticillate so forming apparent terminal verticillasters 1–15 cm long; pedicel 0.7–1 mm long, puberulous, with 1 bract and 2 bracteoles, these sub-verticillate, subequal, ± 2 mm long; sepals linear or narrowly elliptic, 1.5–2.2 mm long, 0.3–0.5 mm wide, ciliate with many hairs similar to those of stem. Corolla pink or mauve, sub-cylindric to narrowly infundibuliform, 2.2–3.8 mm long, the four lobes triangular to ovate, 0.3–0.6 mm long, rounded or obtuse, glabrous and papillose. Filaments 1.3–2 mm long, anthers included, 0.4–0.5 mm long, the thecae slightly divergent, the caudate tails absent or up to 0.2(–0.5) mm long. Ovary pubescent; style 1.3–2 mm long, stigma slightly capitate and 0.2–0.3 mm in diameter, included or slightly exserted. Fruit globose or ovoid, to 2 mm, puberulous, splitting the persistent corolla at maturity.

UGANDA. Elgon, Madangi, Sep. 1932, *A.S.Thomas* 606! & without specific location, Oct. 1996, *Wesche* 124! & Jan. 1997, *Wesche* 854!
KENYA. Cherangani Hills, Chepkotet, Aug. 1968, *Thulin & Tidigs* 257!; Nyandarua/Aberdare Mts, Kinangop, July 1948, *Hedberg* 1662!; Mt Kenya, NW slopes, Mar. 1968, *Mwangangi & Fosberg* 577!
TANZANIA. Kilimanjaro, slopes of Shira Cone, Sep. 1993, *Grimshaw* 93/610!; Arusha District: Mt Meru, W slopes above Olkalau, Oct. 1948, *Hedberg* 2353!; Iringa District: Mt Image, Mar. 1962, *Polhill & Paulo* 1626!
DISTR. **U** 3; **K** 3, 4; **T** 2, 7; N Malawi (Nyika Plateau)
HAB. Afro-alpine moorland, usually on dry and rocky ground in the heath zone and grassland; (2700–)3000–4350 m
USES. None recorded
CONSERVATION NOTES. Least concern (LC) as it occurs in a common habitat range, over a reasonable altitude range

SYN. *Blaeria filago* Alm & T.C.E.Fr. in N.B.G.B. 8: 691 (1924) & in Acta Hort. Berg. 8 (8): 253, t. 10 a, c & 13 f–g (1924); A.V.P.: 147, 306 (1957); R. Ross in F.Z. 7(1): 173 (1983); Blundell, Wild Flowers E.A.: t. 757 (1987); U.K.W.F. ed. 2: 170 (1994)
   B. afromontana Alm & T.C.E.Fr. in N.B.G.B. 8: 692 (1924). Type: Kenya, Kenya, Mt Kenya, *Fries & Fries* 1380 (UPS, holo.; B†, K!, iso.)
   B. viscosa Alm & T.C.E.Fr. in N.B.G.B. 8: 692 (1924) & in Acta Hort. Berg. 8 (8): 254 (1924). Type: Kenya, Nyandarua/Aberdare Mts, *Fries & Fries* 2526 (UPS, lecto., B†, K!, iso.), chosen by Hedberg
   B. elgonensis Alm & T.C.E.Fr. in N.B.G.B. 8: 693 (1924). Type: Kenya, Elgon, *Dummer* 3355 (UPS, holo.; B†, K!, iso.)
   B. filago Alm & T.C.E.Fr. var. *afromontana* (Alm & T.C.E.Fr.) Alm & T.C.E.Fr. in Acta Hort. Berg. 8: 253 (1924)
   B. viscosa Alm & T.C.E.Fr. var. *elgonensis* (Alm & T.C.E.Fr.) Alm & T.C.E.Fr. in Acta Hort. Berg. 8: 254 (1924)
   B. saxicola Alm & T.C.E.Fr. in Acta Hort. Berg. 8: 252, t. 9a & 13e (1924); T.T.C.L.: 191 (1949). Type: Tanzania, Kilimanjaro, Mawenzi, *Volkens* 1366 (B†, holo.; K!, neo., BM!, BR, iso., chosen by Hedberg)
   B. filago Alm & T.C.E.Fr. subsp. *saxicola* (Alm & T.C.E.Fr.) Hedb., A.V.P.: 148, 306 (1957)

NOTE. *B. saxicola* was seen as a distinct subspecies by Hedberg based on slightly smaller corollas (2.5 mm rather than 2.6 mm and larger) and more robust stems.

The protologue states this is close to *B. meyeri-johannis* and *B. saxicola* (both = *Erica silvatica*), "but well distinct". It is not very distinct but there are differences in habit which make me reluctant to lump this with the bulk of *E. silvatica*. However, to write a working key for the two taxa is difficult!

MATERIAL OF UNCLEAR STANDING

A specimen from **T** 7, Njombe District: just off Elton Plateau, Lumakarya stream, May 1953, *Eggeling* 6612! does not key out to any species. It has simple hairs on the stems; 3-nate glabrous leaves; unequal rather broad sepals, a 4-nate corolla, ± 7 anthers, a small capitate stigma, and a glabrous ovary. It is a shrub of 1.8 m from heath at a tream edge at 2400 m. It comes closest to *E. kingaensis* but with some differences – but not enough to describe it as a taxon of its own.

A specimen from **T** 8, Lindi District: Rondo Plateau, St. Cyprian's College, June 2001, *Anthony* 33! is composed of two small branches with leaves in fours; one branch has dendritic hairs on the stem and very few minute hairs on the leaves; the other branch has simple hairs only. The specimen was collected in regenerating thicket at an altitude of 650 m. It is not *Erica mafiensis*; and its altitude is very low, in fact very much below the range of any of the other east African taxa; nor does it resemble any of the F.Z. area taxa. Fertile material would be very interesting!

TAXA OF UNCLEAR STANDING IN OUR AREA

Erica johnstoniana Britten in Trans. Linn. Soc. Lond. ser. 2, Bot. 4: 23, t 5/1–6 (1894); P.O.A. C: 302 (1895); Alm & T.C.E.Fr. in Ark. Bot. 21a (7): 20 (1927); Weimarck in Bot. Not. 1940: 54 (1940); Brenan in Mem. N.Y. Bot. Gard. 8: 493 (1954); R. Ross in F.Z. 7(1): 169 (1983). Type: Malawi, Mlanje Plateau, *Whyte* 4 (BM, holo.; K, iso.)

Occurs in Tanzania according to FZ footnote, p. 166 – a specimen said to be from Tanzania, *Swynnerton* 2015 (BM!). Identification is certainly correct, but I doubt whether it is from Tanzania. There is the usual printed Swynnerton label with "Tanganyika Territory Game Department – C.F.M. Swynnerton" on it, but usually the locality is hand-written in by the collector; here there is nothing. The specimen number is on a loose bit of paper. I think this specimen was probably collected in Malawi, and I do not believe it occurs in the FTEA area.

Erica microdonta (C.H.Wright) E.G.H.Oliv. in Bothalia 24: 124 (1925), synonym *Ericinella microdonta* (C.H.Wright) Alm & T.C.E.Fr. in Acta Hort. Berg. 8 (8): 262 (1924) & in K. Svenska Vetensk.-Akad. Handl. ser. 3, 4 (4): 46 (1927); Brenan in Mem. N.Y. Bot Gard. 8: 494 (1954); R. Ross in F.Z. 7(1): 174, t. 30 (1983). Type: Malawi, Mt Mlanje, *McClounie* 55, 75, 95 (K, syn.)

Ross in F.Z. says this also occurs in SW Tanzania, but I have seen no specimens.

EXCLUDED TAXA

Erica lanceolifera S.Moore in J.L.S. 40: 126 (1911); Alm & T.C.E.Fr. in Ark. Bot. 21a (7): 8 (1927); Brenan in Kirkia 4: 146 (1964); R. Ross in F.Z. 7(1): 166 (1983). Type: Zimbabwe, Chimanimani Mts, *Swynnerton* 1288 (BM, holo.; K, iso.)

"Possibly in Tanzania" according to FZ – based on a doubtful specimen from Kilimanjaro at 3000 m, collected by *Swynnerton* (numbered 1040!). This taxon (*lanceolifera*) occurs within Swynnerton's normal collecting area in Zimbabwe; it has never been found by anyone else on Kilimanjaro or in other parts of Tanzania; I believe it is a case of mislabelling, and the specimen hails from Zimbabwe.

INDEX TO ERICACEAE

New names validated in this part

Erica trimera (*Engl.*) *Beentje* subsp. **jaegeri** (*Engl.*) *Beentje* **comb. nov**.
Erica trimera (*Engl.*) *Beentje* subsp. **kilimanjarica** (*Engl.*) *Beentje* **comb. nov**.
Erica silvatica (*Engl.*) *Beentje* **comb. nov**.
Erica filago (*Alm. & T.C.E.Fr.*) *Beentje* **comb. nov**.

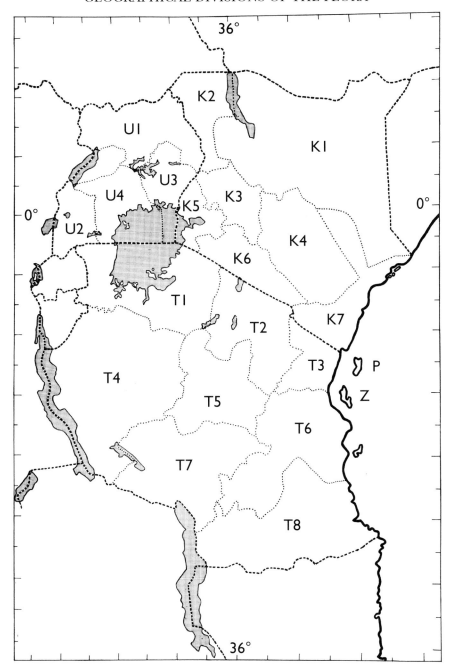

PLANTS PEOPLE
POSSIBILITIES

First published in 2006 by
Royal Botanic Gardens, Kew
Richmond, Surrey, TW9 3AB, UK
www.kew.org

ISBN 1 84246 144 3

Design by Media Resources, typesetting and page layout by Margaret Newman,
Information Services Department,
Royal Botanic Gardens, Kew.

Printed in the UK by Hobbs the Printers

For information or to purchase all Kew titles please visit
www.kewbooks.com or email publishing@kew.org